71
Science
Experiments

Making Science Simpler For You

Vikas Khatri

V&S PUBLISHERS

Published by:

V&S PUBLISHERS

F-2/16, Ansari Road, Daryaganj, New Delhi-110002
011-23240026, 011-23240027• *Fax:* 011-23240028
Email: info@vspublishers.com • *Website:* www.vspublishers.com

Branch : Hyderabad
5-1-707/1, Brij Bhawan (Beside Central Bank of India Lane)
Bank Street, Koti, Hyderabad - 500 095
040-24737290
E-mail: vspublishershyd@gmail.com

Follow us on: t f in

For any assistance sms **VSPUB** to **56161**

All books available at **www.vspublishers.com**

© **Copyright:** *V&S PUBLISHERS*
ISBN 978-93-815884-6-8
Edition 2013

Printed at : Param Offsetters, Okhla, New Delhi

Publisher's Note

If you are looking for some incredible Science Experiments for your children, then you have come to the right place and chosen the right book. Check out for some fun-filled, absolutely novel and informative experiments divided into small chapters which will make it easier to grasp. They are full of fascinating ideas and facts that have been implemented in a interesting way for children of all ages to enjoy the world of science. Learn the amazing Science and Technology facts as you experiment with different common materials used in our day to day lives, found in and around our houses that react in all sorts of mesmerising and crazy ways.

However, all these experiments should be performed carefully under adult supervision only. These science projects are knowledgeable, safe and easy to perform.

So, enjoy carrying out the given incredible science experiments in this book.

- Make cool science projects with novel ideas.
- Learn amazing science and technology facts.
- Show your friends and family what you have discovered.

Last but not the least a word for the **Teachers**. This book is an ideal one and serves as a true science guide. It acts as a handbook for teachers, particularly teaching the science subjects, as Science can never be *understood* and *simplified* without experiments.

So all school going children can buy it and make best use of it. It can also be handy for teachers, while teaching the children the various laws, discoveries, inventions and scientific phenomena, make easy through these 71 incredible experiments compiled in the book.

Read on and enjoy the thrill of performing these experiments that enrich your scientific knowledge making *Science simpler for you, as never before.*

Contents

Introduction

Science, as we all know is a subject which is universal and knows no boundaries. It is also a pivotal subject in the school curricula of almost all the schools, across the globe. Hence, there is no need to explain the significance of science for students. It is invaluable and indispensable in today's world.

However, science can never be taught theoretically. It always needs practical demostrations through simple experiments and projects as illustrated in this book. There are around 71 interesting experiments explaining the various simple scientific phenomena of gravity, temperature changes, air pressure, boiling, melting and freezing points of solids and liquids, magnetism, electricity, etc., all compiled together in easy language for the children to grasp quickly, and carry them out in schools and homes with adults' or teachers' supervision, of course.

The benefits of learning science through experiments are that children do not just understand the concepts but also implement them practically, clearing and testing their ideas independently. They are able to evaluate themselves and use their knowledge in future – not just confine themselves to classroom studies because Experimentation is a must in Scientific Studies to methodically verify, falsify or establish the validity and truth of an invention, theory or law.

Remember to always have fun with your science experiments and don't worry if your experiments don't come out as you would expect – some of the greatest scientific discoveries have been made by mistake.

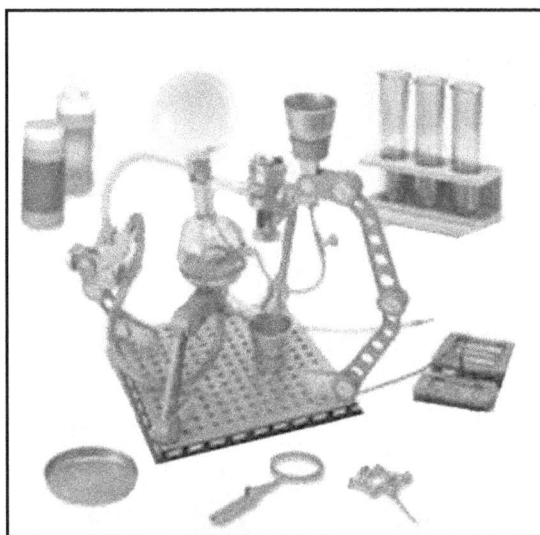

Science is all around us in our daily living and the more you experiment with science and observe; the more fascinated you will become in finding answers.

Hence, go ahead, read the given fun-filled experiments and perform them at home or school with the help of your parents or teachers. Thus, you can also enjoy while you learn.

Make an Egg Float in Salt Water

An egg sinks to the bottom if you drop it into a glass of ordinary drinking water. But what happens if you add salt? The results are very interesting and can teach you some facts about density.

What you need

One egg, Water, Salt, A tall drinking glass.

SALT

What to do

1. Pour water into the glass until it is about half full.
2. Stir in lots of salt (about 6 tablespoons).

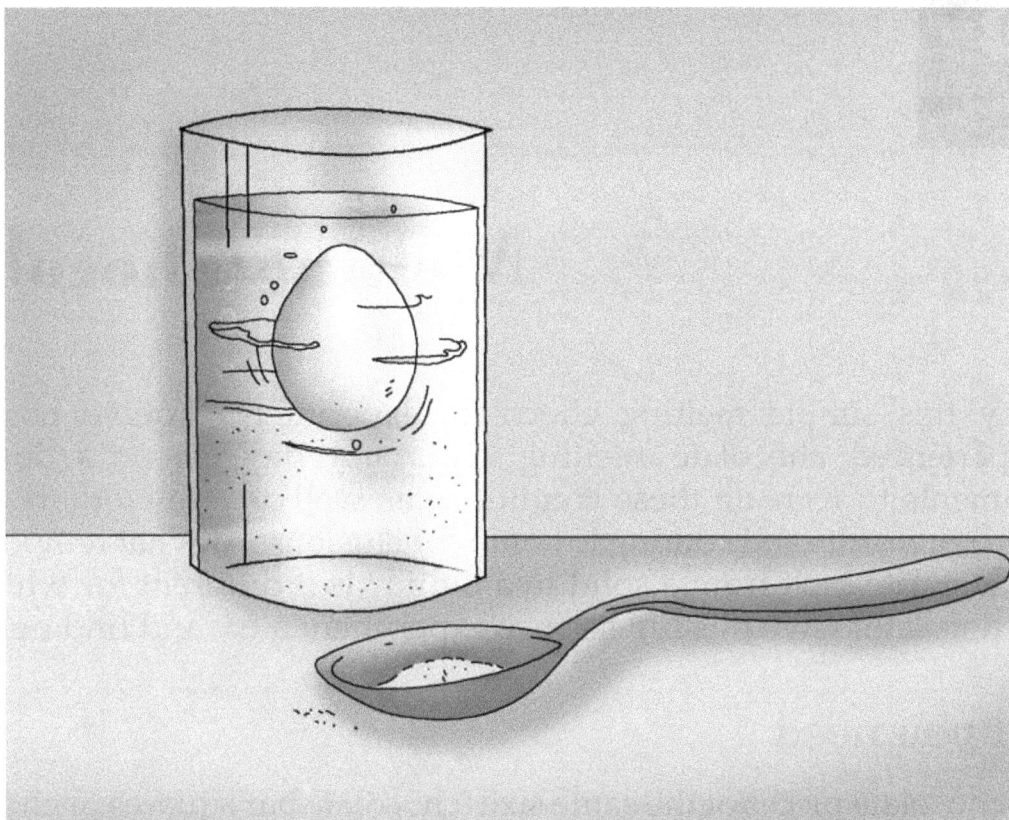

3. Carefully pour in plain water until the glass is nearly full. (Be careful not to disturb or mix the salty water with the plain water.)

4. Gently lower the egg into the water and watch what happens.

What's happening

Salt water is denser than ordinary tap water; the denser the liquid, the easier it is for an object to float in it. When you lower the egg into the liquid, it drops through the normal tap water until it reaches the salty water. At this point the water is dense enough for the egg to float. If you were careful when you added the tap water to the salt water, they would not have got mixed, enabling the egg to amazingly float in the middle of the glass.

Melting Chocolate

Enjoy this simple melting chocolate experiment. You've no doubt experienced chocolate melting on a hot day, so let's do some experiments to recreate these conditions as well as a few others before comparing results and coming to some conclusions. At what temperature does chocolate melt from a solid to a liquid? Is it different for white and dark chocolate? Give this fun science experiment a try and find out!

What you need

Small chocolate pieces of the same size (chocolate bar squares or chocolate chips are a good idea), Paper plates, Pen and paper to record your results.

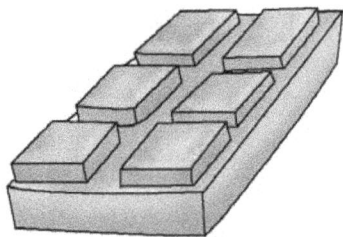

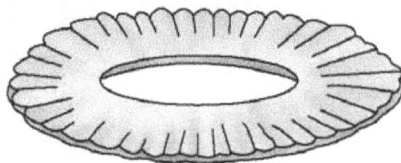

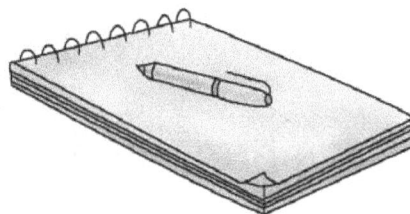

What to do

1. Put one piece of chocolate on a paper plate and put it outside in the shade.
2. Record how long it took for the chocolate to melt, or if it wasn't hot enough to melt then record how soft it was after 10 minutes.
3. Repeat the process with a piece of chocolate on a plate that you put outside in the sun. Record your results in the same way.

4. Find more interesting locations to test how long it takes for the chocolate pieces to melt. You could try your school bag, hot water or even your own mouth.

5. Compare your results, in what conditions did the chocolate melt? You might also like to record the temperatures of the locations you used using a thermometer. Thus you can find out the exact temperature at which the chocolate melts.

What's happening

At a certain temperature, your chocolate pieces undergo a physical change, from a solid to a liquid (or somewhere in between). On a hot day, sunlight is usually enough to melt chocolate, something you might have unfortunately already experienced. You can also reverse the process by putting the melted chocolate into a fridge or freezer, where it will go from a liquid back to a solid. The chocolate probably melted quite fast if you tried putting a piece in your mouth. What does this tell you about the temperature of your body? For further testing and experiments, you could compare white chocolate and dark chocolate. Do they melt at the same temperature? How about putting a sheet of aluminium foil between a paper plate and a piece of chocolate in the sun, what happens then?

Mixing Oil & Water

Some things just don't get along well with each other. Take oil and water as an example; you can mix them together and shake as hard as you like, but they'll never become friends.....or will they? Take this fun experiment a step further and find out how bringing oil and water together can help you wash your dishes and keep them sparkling clean.

What you need

Small soft drink bottle, Water, Food colouring, 2 tablespoons of cooking oil and Dish washing liquid or detergent.

What to do

1. Add a few drops of food colouring to the water.
2. Pour about 2 tablespoons of the coloured water along with the 2 tablespoons of cooking oil into the small soft drink bottle.
3. Screw the lid of the bottle tightly and shake the bottle as hard as you can.

4. Put the bottle back down and have a look, it may have seemed as though the liquids were mixing together but the oil will float back to the top.

What's happening

While water often mixes with other liquids to form solutions, oil and water do not mix together. Water molecules are strongly attracted to each other, this is the same for oil. Because they are more attracted to their own molecules. They just don't mix together. They separate and the oil floats above the water because it has a lower density.

If you really think oil and water belong together, then try adding some dish washing liquid or detergent. Detergent is attracted to both water and oil helping them all join together and form something called an *emulsion*. This is extra handy when washing those greasy dishes. The detergent takes the oil and grime off the plates and into the water, yay!

Make your Own Quick Sand

Quick sand is a fascinating substance. Make some of your own and experiment on a safe scale. Surprise your friends and family members by demonstrating how it works.

What you need

1 cup of maize cornflour, Half a cup of water, A large plastic container and A spoon.

What to do

1. This one is simple. Just mix the cornflour and water thoroughly in the container to make your own instant quick sand.
2. When showing other people how it works, stir slowly and drop the quick sand to show it is a liquid.

3. Stirring it quickly will make it hard and allow you to punch or poke it quickly. (This works better if you do it fast rather than hard).
4. Remember that quick sand is messy, try to play with it outside and don't forget to stir just before you use it.
5. Always stir instant quicksand just before you use it!

What's happening

If you add just the right amount of water to cornflour, it becomes very thick when you stir it quickly. This happens because the cornflour grains are mixed up and can't slide over each other due to the lack of water between them. Stirring slowly allows more water between the cornflour grains, letting them slide over each other easily.

Poking it quickly has the same effect, making the substance very hard. If you poke it slowly, it doesn't mix up the mixture in the same way, leaving it runny. It works in much the same way as real quick sand.

Baking Soda & Vinegar Volcano

Use baking soda and vinegar to create an awesome chemical reaction! Watch as it rapidly fizzes over the container and make sure you've got some towels ready to clean up.

What you need

Baking Soda (make sure it's not baking powder), Vinegar, A container to hold everything and avoid a big mess and Paper towels or a cloth (just in case.)

What to do

1. Put some of the baking soda into your container.
2. Pour in some of the vinegar.
3. Watch as the reaction takes place!

What's happening

The baking soda (sodium bicarbonate) is a base while the vinegar (acetic acid) is an acid. When they react together, they form carbonic acid which is very unstable. It instantly breaks apart into water and carbon dioxide, which creates all the fizzing as it escapes the solution.

For extra effect, you can make a realistic looking volcano. It takes some craft skills but it will make your vinegar and baking soda eruptions look even more impressive!

Raw or Boiled Egg?

Surprise your friends and family with an easy science experiment that answers a tricky question. Two eggs look and feel the same, but there is a big difference– one is raw and the other, hard boiled. Find out which is which with this fun experiment.

What you need

Two eggs, one hard boiled and one raw. Make sure the hard boiled egg has been in the fridge long enough to be of the same temperature as the raw egg.

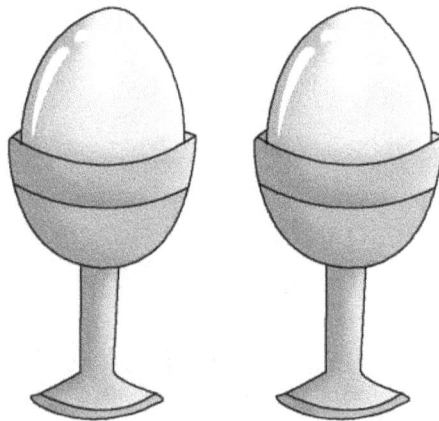

What to do

1. Spin the eggs and watch what happens– one egg should spin while the other wobbles.
2. You can also lightly touch each of the eggs while they are spinning. One should stop quickly while the other keeps moving after you have touched it.

What's happening

The raw egg's centre of gravity changes as the white and yolk move around inside the shell, causing the wobbling motion. Even after you touch the shell, it continues moving. This is because of *inertia*. The same type of force is felt when you change direction or stop suddenly in a car, your body wants to move one way while the car wants to do something different. Inertia causes the raw egg to spin even after you have stopped it. This contrasts with the solid white and yolk of the hard boiled egg, it responds much quicker if you touch it.

This is a good experiment to test a friend or someone in your family, and see if they can figure out the difference between the two eggs (without smashing them, of course) before showing them your nifty trick.

Make Glowing Water

Make glowing water with the help of a black light in this fun science experiment for kids. Tonic water doesn't look very strange under normal light but what happens when you look at it under a black light? Does the dye from a highlighter pen do the same thing? Find out what happens and why it happens with this cool experiment that you can do at home.

What you need

A black light lamp, Tonic water or a highlighter pen and A dark room to do the experiment.

What to do

1. If you are using a highlighter pen, carefully break it open, remove the felt and soak it in a small amount of water for a few minutes.
2. Find a dark room.
3. Turn on the black light near your water. How does it look?

What's happening

The ultraviolet (UV) light coming from your black light lamp excites the things called phosphors. Tonic water and the dye from the highlighter pens contain phosphors that turn the UV light (light we can't see) into visible light (light we can see). That's why your water glows in the dark when you a black light on it.

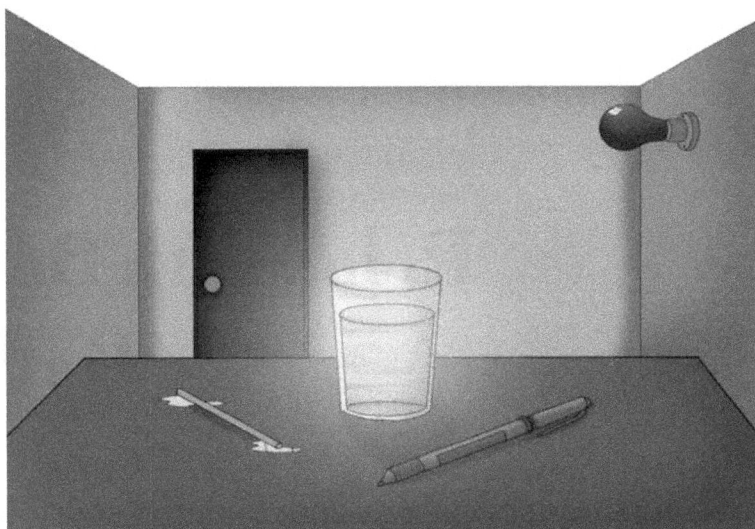

Black lights are used in forensic science, artistic performances, photography, authentication of banknotes and antiques, and in many other areas.

What's more

Black light (also known as the UV or the ultraviolet light) is a part of the electromagnetic spectrum. The electromagnetic spectrum also includes infrared, X-rays, visible light (what the human eye can see) and other types of electromagnetic radiation. A black light lamp, such as the one you used, emits a UV light that can illuminate objects and materials that contain phosphors. Phosphors are special substances that emit light (luminescence) when excited by radiation. Your water glowed under the black light because it contained phosphors. If you had used a highlighter pen, then the UV light would have reacted with the phosphors in the dye. If you had used tonic water, then the UV light reacted with the phosphors in the chemical used in tonic water called quinine.

There are different types of luminescence. These include fluorescence (used in this experiment, it glows only when the black light is on), phosphorescence (similar to fluorescence but with a glow that can last even after the black light is turned off), chemiluminescence (used to create glow sticks), bioluminescence (from living organisms) and many others.

Relax with Beautiful Bath Salts

Whether you're making a special present for someone else, experimenting at home or just want to relax in a hot bath, give this experiment a try. Create your own bath salts with a variety of refreshing fragrances, and experiment with different essential oils to see which you like most.

What you need

1 cup of washing soda, A plastic bag, A rolling pin (or something similar that can crush lumps), A bowl, A spoon for stirring, Essential oil and Food colouring.

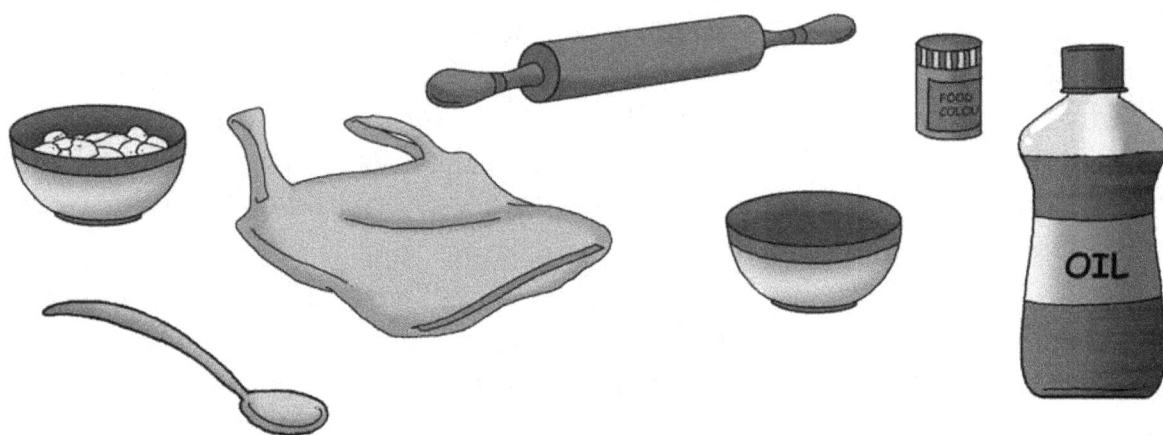

What to do

1. Take the cup of washing soda and put it into a plastic bag. Crush the lumps with a rolling pin or similar object.

2. Empty the bag into a bowl and stir in 5 or 6 drops of your favourite essential oil, such as rosemary, lavender or mint.
3. Stir in a few drops of food colouring until the mixture is evenly coloured.
4. Put the mixture into clean dry containers and enjoy as you please.

What's happening

Bath salts are typically made from Epsom salts (magnesium sulphate), table salt (sodium chloride) or washing soda (sodium carbonate). The chemical make up of the mixture makes it easy to form a lather. Bath salts are said to improve cleaning and deliver an appealing fragrance when bathing.

Grow your Own Bacteria

Bacteria are a fascinating type of microorganisms that play a vital role in our lives whether we like it or not. Try growing your own sample of bacteria while monitoring how it reproduces in a short span of time. Compare your original sample with others and get a proof that bacteria truly are everywhere!

What you need

Petri dish of agar, Cotton buds, Some old newspaper (to wrap the petri dish when disposing)

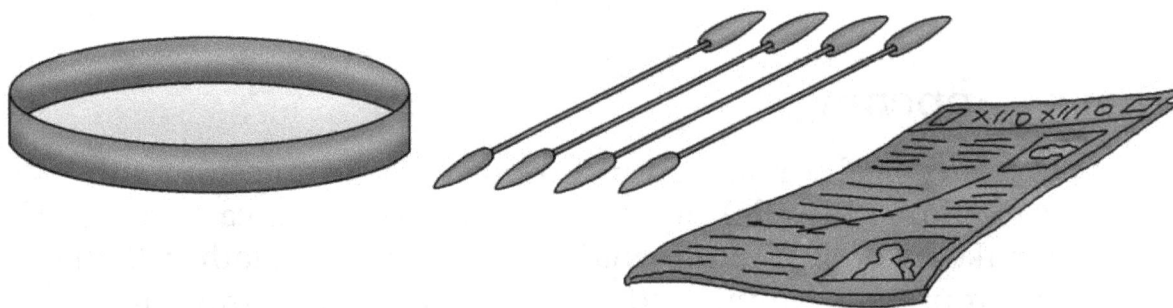

What to do

1. Prepare your petri dish of agar.
2. Using your cotton bud, swab a certain area of your house (i.e. collect a sample by rubbing the cotton bud on a surface of your choice).
3. Rub the swab over the agar with a few gentle strokes before putting the lid back on and sealing the petri dish.
4. Allow the dish to sit in a warm area for 2 or 3 days.

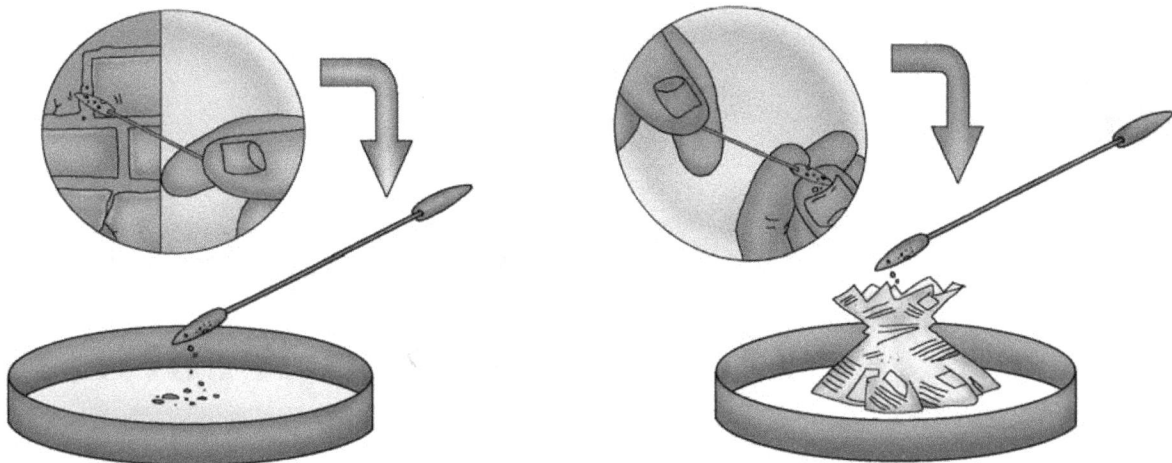

5. Check the growth of the bacteria each day by making an observational drawing and describing the changes.
6. Try repeating the process with a new petri dish and swab from under your finger nails or between your toes.
7. Dispose of the bacteria by wrapping up the petri dish in an old newspaper and placing in the rubbish (don't open the lid).

What's happening

The agar plate and warm conditions provide the ideal place for the bacteria to grow. The microorganisms on the plate will grow into individual colonies, each a clone of the original. The bacteria you obtained with the cotton bud grows steadily, becoming visible with the naked eye in a relatively short time. Different samples produce different results. What happened when you took a swab sample from your own body?

You will find bacteria throughout the earth, it grows in soil, radioactive waste, water, on plants and even animals too (humans included). Thankfully for us, our immune system usually does a great job of making bacteria harmless.

Dissolving Sugar at Different Temperatures

Learn about solutions as you add more and more sugar cubes to different temperatures of water. This easy experiment shows that you can only dissolve a certain amount and that the amount changes as the water gets hotter.

What you need

Sugar cubes, Cold water in a clear glass, Hot water in a clear glass (be careful with the hot water) and Spoon for stirring.

What to do

1. Make sure the glasses have an equal amount of water.
2. Put a sugar cube into the cold water and stir with the spoon until the sugar disappears. Repeat this process (remembering to count the amount of sugar cubes you put into the water) until the sugar stops dissolving. At this point, the sugar starts to gather on the bottom of the glass rather than dissolving.
3. Write down how many sugar cubes you could dissolve in the cold water.
4. Repeat the same process with the hot water, and compare the number of sugar cubes dissolved in each liquid. Which dissolved more?

What's happening

The cold water isn't able to dissolve as much sugar as the hot water, but why? Another name for the liquids inside the cups is 'solution'. When this solution can no longer dissolve sugar, it becomes a 'saturated solution'. This means that sugar starts forming on the bottom of the cup.

The reason that hot water dissolves more is because it has faster moving molecules which are spread further apart than the molecules in the cold water. With bigger gaps between the molecules in the hot water, more sugar molecules can fit in between.

Making Music with Water

Have you ever tried making music with glasses or bottles filled with water? I bet you haven't. Experiment with your own special sounds by turning glasses of water into instruments, make some cool music and find out how it works.

What you need

5 or more drinking glasses or glass bottles, Water and a Wooden stick or a pencil.

What to do

1. Line the glasses up next to each other and fill them with different amounts of water. The first should have just a little water, while

the last should almost be full, and the ones in between should have slightly more than the last.

2. Hit the glass with the least amount of water and observe the sound, then hit the glass with the most water. Which one makes the higher sound?

3. Hit the other glasses and see what noise they make, see if you can get a tune going by hitting the glasses in a certain order.

What's happening

Each of the glasses will have a different tone when hit with a pencil or a wooden stick. The glass with the most water will have the lowest tone while the glass with the least water will have the highest. Small vibrations are made when you hit the glass. This creates sound waves which travel through the water. More water means slower vibrations and a deeper tone.

Use a Balloon to Amplify Sound

Small sounds can still make a big noise when you use a good sound conductor. Experiment with a balloon, compressed air and your own ears to find out how it works.

What you need

A balloon

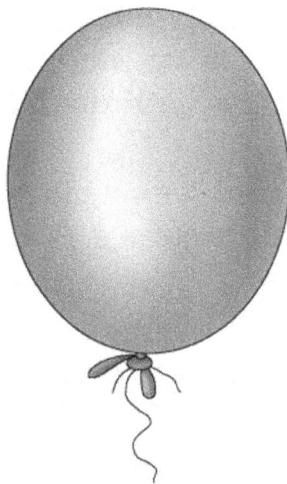

What to do

1. Blow up the balloon.
2. Hold the balloon close to your ear while you tap lightly on the other side.

What's happening

Despite your tapping lightly on the balloon, your ears can hear the noise loudly. When you blew up the balloon, you forced the air molecules inside the balloon closer to each other. Because the air molecules inside the balloon are closer together, they become a better conductor of sound waves than the ordinary air around you.

Make a Ping Pong Ball Float

Can you control a ping pong ball as it floats above a hair dryer? Put your hand-eye coordination skills to the test while learning the important role that forces such as gravity and air pressure play in this simple experiment for kids.

What you need

At least 1 ping pong ball (2 or 3 would be great) and A hair dryer.

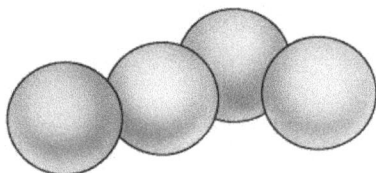

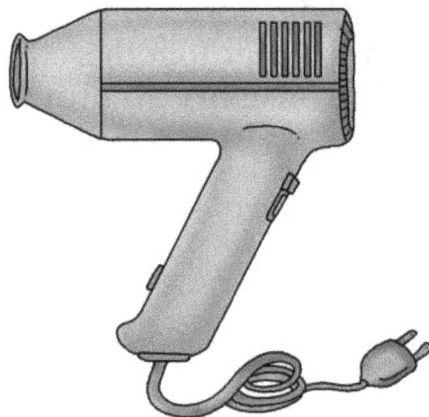

What to do

1. Plug in the hair dryer and turn it on.
2. Put it on the highest setting and point it straight up.
3. Place your ping pong ball above the hair dryer and watch what happens.

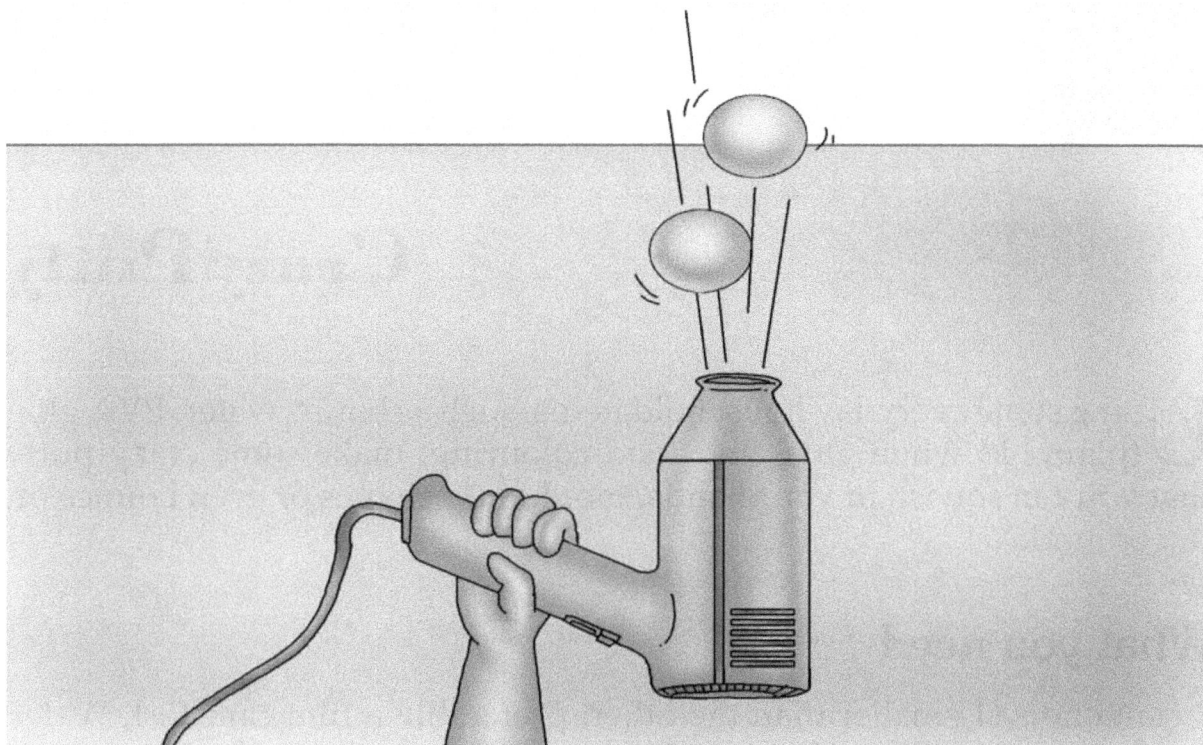

What's happening

Your ping pong ball floats gently above the hair dryer without shifting sideways or flying across the other side of the room. The airflow from the hair dryer pushes the ping pong ball upwards until its upward force equals the force of gravity pushing down on it. When it reaches this point, it gently bounces around, floating where the upward and downward forces are equal.

The reason the ping pong ball stays nicely inside the column of air produced by the hair dryer without shifting sideways is due to air pressure. The fast moving air from the hair dryer creates a column of lower air pressure, while the surrounding higher air pressure forces the ping pong ball to stay inside this column, making it easy to move the hair dryer around without losing control of the ping pong ball.

See if you can float 2 or even 3 ping pong balls as an extra challenge.

Crazy Putty

Using some everyday household items, such as borax, water, PVA glue (plain old while glue) and food colouring, make some crazy putty that you can squash in your hands, mould into shapes or even bounce on the ground.

What you need

2 containers (1 smaller than the other, preferably a film canister), Water, Food colouring, PVA glue and Borax solution (ratio of about 1 tbsp of borax to a cup of water.)

What to do

1. Fill the bottom of the larger container with PVA glue.
2. Add a few squirts of water and stir.
3. Add 2 or 3 drops of food colouring and stir.
4. Add a squirt of borax (possibly a bit more depending on how much PVA glue you have used).
5. Stir the mixture up and put it into the smaller container. By now the mixture should be joining together, acting like putty, crazy putty!

What's happening

The PVA glue you use is a type of polymer called Polyvinyl Acetate (PVA in short), while the borax is made of a chemical called Sodium Borate. When you combine the two in a water solution, the borax reacts with the glue molecules, joining them together into one giant molecule. This new compound is able to absorb large amounts of water, producing a putty like substance which you can squash in your hands or even bounce.

Experience Gravity Free Water

What go up must come down, right? Well try bending the rules a little with a cup of water that stays inside the glass when held upside down. You'll need the help of some cardboard and a little bit of air pressure.

What you need

A glass filled right to the top with water and a piece of cardboard.

What to do

1. Put the cardboard over the mouth of the glass, making sure that no air bubbles enter the glass as you hold on to the cardboard.

2. Turn the glass upside down (over a sink or outside).
3. Take away your hand holding the cardboard.

What's happening

If all goes according to the plan, then the cardboard and water should stay put. Even though the cup of water is upside down, the water stays in place, defying gravity! So why is this happening? With no air inside the glass, the air pressure from outside the glass is greater than the pressure of the water inside the glass. The extra air pressure manages to hold the cardboard in place, keeping you dry and your water where it should be, inside the glass.

What is your Lung Volume?

Do you think you're fit and healthy? Let's find out your lung volume. Just how much air can your lungs hold? With the help of a few simple household objects, some scientific know-how and a dash of curiosity, you can make this experiment look easy.

What you need

Clean plastic tubing, A large plastic bottle, Water and A Kitchen sink or a large water basin.

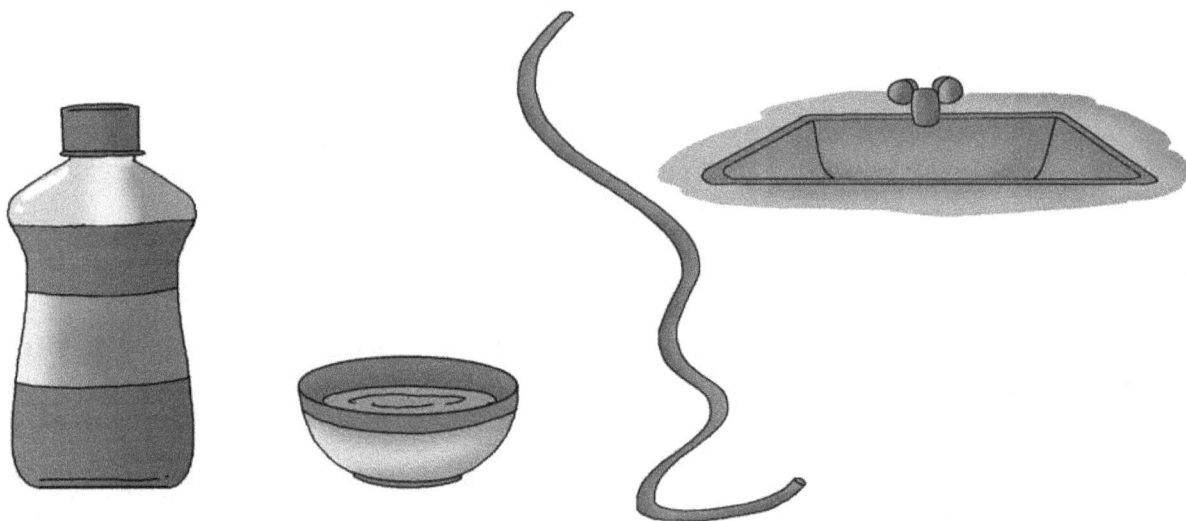

What to do

1. Make sure the plastic tubing is clean.
2. Put about 10cm of water into your kitchen sink.
3. Fill the plastic bottle right to the top with water.
4. Put your hand over the top of the bottle to stop the water from escaping when you turn it upside down.

5. Turn the bottle upside down. Place the top of the bottle under the water in the sink before removing your hand.
6. Push one end of the plastic tube into the bottle.
7. Take a big breath in.
8. Breathe out as much air as you can through the tube.
9. Measure the volume of air your lungs had in them.
10. Make sure you clean up the area to finish.

What's happening

As you breathe out through the tube, the air from your lungs takes the place of the water in the bottle. If you made sure you took a deep breath in and breathed out fully, then the resulting volume of water you pushed out is equivalent to the volume of air your lungs can hold. Having a big air capacity in your lungs means you can distribute oxygen around your body at a faster rate. The air capacity of lungs (or VO2 max) increases naturally as children grow up, but can also be increased with regular exercise.

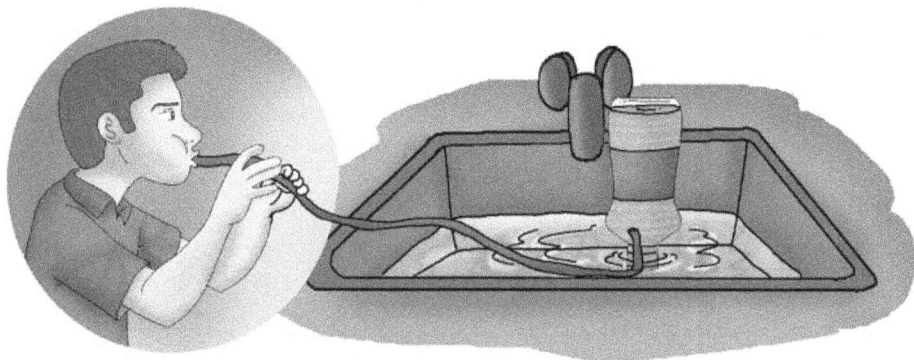

Make a Crystal Snowflake!

Learn how to make a snowflake by using borax and a few other easy-to-find household items. Find out how crystals are formed in this fun crystal activity, experiment with food colouring to enhance the look and keep your finished crystal snowflake as a great looking decoration!

What you need

A String, A wide-mouthed jar, White pipe cleaners, Blue food colouring (optional), Boiling water (take care or better still get an adult to help), Borax, A small wooden rod or pencil.

What to do

1. Grab a white pipe cleaner and cut it into three sections of the same size. Twist these sections together in the centre so that you now have a shape that looks something like a six-sided star. Make sure the points of your shape are even by trimming them to the same length.
2. Take the top of one of the pipe cleaners and attach another piece of string to it. Tie the opposite end to your small wooden rod or pencil. You will use this to hang your completed snowflake.
3. Carefully fill the jar with boiling water (you might want to get an adult to help with this part).

4. For each cup of water, add three tablespoons of borax, adding one tablespoon at a time. Stir until the mixture is dissolved but don't worry if some of the borax settles at the base of the jar.

5. Add some of the optional blue food colouring if you'd like to give your snowflake a nice bluish tinge.

6. Put the pipe cleaner snowflake into the jar so that the small wooden rod or pencil is resting on the edge of the jar and the snowflake is sitting freely in the borax solution.

7. Leave the snowflake overnight and when you return in the morning, you will find the snowflake covered with crystals! It makes a great decoration that you can show to your teachers and friends or hang somewhere in your house.

What's happening

Crystals are made up of molecules arranged in a repeating pattern that extends in all three dimensions. Borax is also known as sodium borate. It is usually found in the form of a white powder made up of colourless crystals that are easily dissolved in water.

When you add the borax to the boiling water, you can dissolve more than you could if you were adding it to cold water. This is because warm water molecules move around faster and are more spread apart, allowing more room for the borax crystals to dissolve.

When the solution cools, the water molecules move closer together and is unable to hold as much of the borax solution. Crystals begin to form on top of each other and before you know it you have your completed crystal snow flake!

Invisible Ink with Lemon Juice

Making invisible ink is a lot of fun. You can pretend that you are a secret agent as you keep all your secret codes and messages hidden from others. All you need is some basic household objects and the hidden power of lemon juice.

What you need

Half a lemon, Water, A spoon, A bowl, Cotton bud, White paper, A lamp or other light bulb.

What to do

1.   Squeeze some lemon juice into the bowl and add a few drops of water.

2. Mix the water and lemon juice with the spoon.
3. Dip the cotton bud into the mixture and write a message onto the white paper.
4. Wait for the juice to dry so that it becomes completely invisible.
5. When you are ready to read your secret message or show it to someone else, heat the paper by holding it close to a light bulb.

What's happening

Lemon juice is an organic substance that oxidizes and turns brown when heated. Diluting the lemon juice in water makes it very hard to notice when you apply it on the paper. No one will be aware of its presence until it is heated and the secret message is revealed. Other substances which work in the same way include orange juice, honey, milk, onion juice, vinegar and wine. Invisible ink can also be made using chemical reactions or by viewing certain liquids under the ultraviolet (UV) light.

Make an Easy Lava Lamp

Learn how to make an easy lava lamp with this fun science experiment for kids. Use simple household items, such as vegetable oil, food colouring, Alka-Seltzer and a bottle to create chemical reactions and funky balls of colour that move around just like a real lava lamp.

What you need

Water, A clear plastic bottle, Vegetable oil, Food colouring, Alka-Seltzer (or other tablets that fizz).

What to do

1. Pour water into the plastic bottle until it is around one quarter full. (You might want to use a funnel when filling the bottle so you don't spill anything).
2. Pour in vegetable oil until the bottle is nearly full.
3. Wait until the oil and water have separated.
4. Add around a dozen drops of food colouring to the bottle. (Choose any colour you like).
5. Watch as the food colouring falls through the oil and mixes with the water.

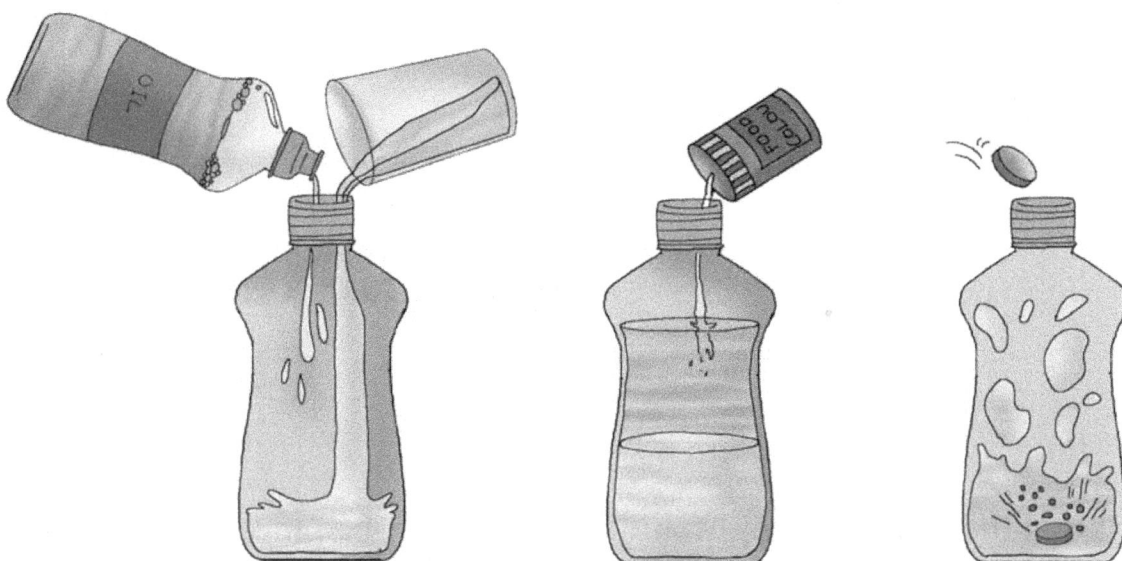

6. Cut an Alka-Seltzer tablet into smaller pieces (around 5 or 6) and drop one of them into the bottle. Things should start getting a little crazy, just like a real lava lamp!

7. When the bubbling stops, add another piece of Alka-Seltzer and enjoy the show!

What's happening

If you've tried our oil and water experiment, you'll know that the two don't mix very well. The oil and water you added to the bottle separate from each other, with oil on top because it has a lower density than water. The food colouring falls through the oil and mixes with the water at the bottom. The piece of Alka-Seltzer tablet you drop in releases small bubbles of carbon dioxide gas that rise to the top and take some of the coloured water along for the ride. The gas escapes when it reaches the top and the coloured water falls back down. The reason Alka-Seltzer fizzes in such a way is because it contains citric acid and baking soda (sodium bicarbonate). The two react with water to form sodium citrate and carbon dioxide gas. (Those are the bubbles that carry the coloured water to the top of the bottle).

Adding more Alka-Seltzer to the bottle keeps the reaction going so that you can enjoy your funky lava lamp for longer. If you want to show someone later, you can simply screw on a bottle cap and add more Alka-Seltzer, when you need to. When you've finished all your Alka-Seltzer, you can take the experiment a step further by tightly screwing on a bottle cap and tipping the bottle back and forth, what happens then? Do it, and find out for yourself.

Will the Ice Melt & Overflow?

At first thought, you might think that an ice cube placed at the very top of a glass would eventually melt and spill over the sides, but is this what really happens? Experiment and find out!

What you need

A clear glass, Warm water, An ice cube.

What to do

1. Fill the glass to the top with warm water.
2. Gently lower in the ice cube, making sure you don't bump the table or spill any water over the edge of the glass.

3. Watch the water level carefully as the ice cube melts, what happens?

What's happening

Even though the ice cube melted, the water doesn't overflow. When water freezes to make ice, it expands and takes up more space than it does as liquid water. (That's why water pipes sometimes burst during cold winters). The water from the ice takes up less space than the ice itself. When the ice cube melts, the level of the water stays about the same. Interesting, isn't it?

Test your Dominant Side

Check out this cool experiment that will teach you more about how your body and brain work together. Test your dominant side by completing a series of challenges.

Which hand do you write with? Which foot do you kick with? Do you have a dominant eye? Do you throw with one side of your body but kick with the other? Are you ambidextrous? Answer these questions and much more with this incredible experiment for kids.

What you need

A pen or pencil, Paper or a notepad to write your findings on, An empty tube (an old paper towel tube is good), A cup of water and A small ball (or something soft you can throw).

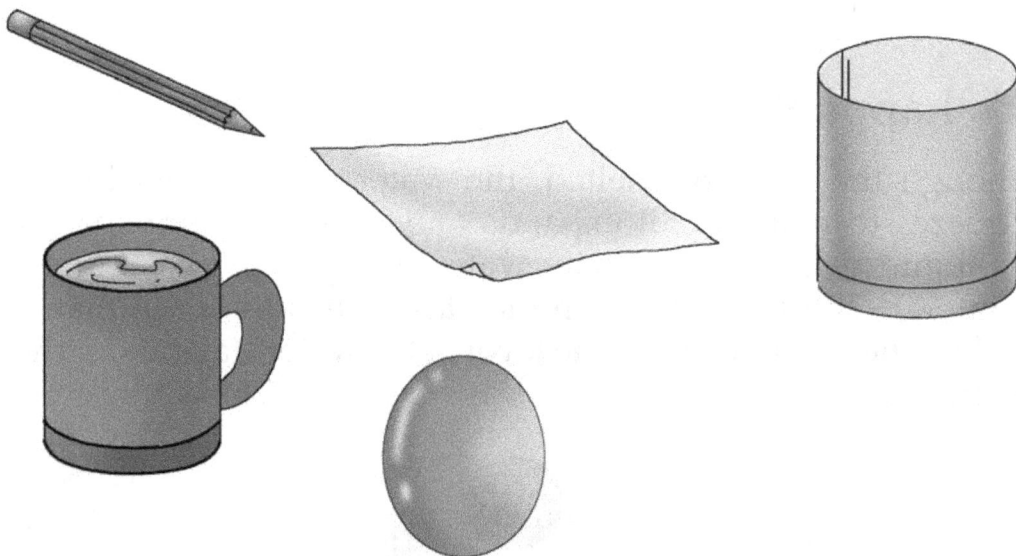

What to do

1. Write 'left' or 'right' next to each task depending on what side you used/favoured.
2. When you've finished all the challenges, review your results and make your own conclusions about which is your dominant eye, hand or foot.

Eye Tests:

1. Which eye do you use to wink?
2. Which eye do you use to look through the empty tube?

3. Extend your arms in front of your body. Make a triangle shape using your forefingers and thumbs. Bring your hands together, making the triangle smaller (about the size of a coin is good). Find a small object in the room and focus on it through the hole in your hands (using both eyes). Try closing just your left eye and then just your right. If your view of the object changed when you closed your left eye, mark down 'left', and if it changed when you closed your right eye, mark down 'right'.

Hand/Arm tests:

1. Which hand do you use to write?
2. Pick up the cup of water, which hand did you use?
3. Throw the ball, which arm did you use?

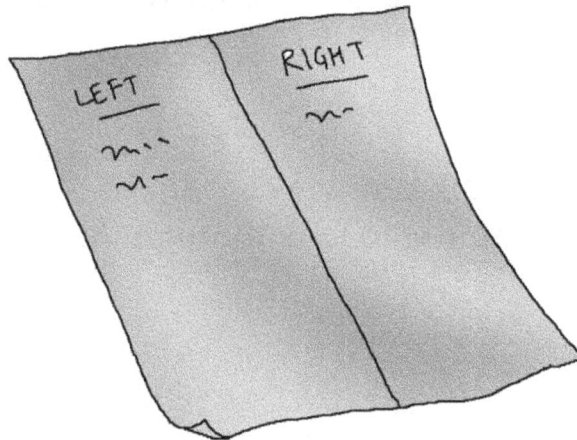

Foot/Leg tests:

1. Run forward and jump off one leg, which did you jump off?
2. Drop the ball on the ground and kick it, which foot did you use?

What's happening

So what side do you favour? Are you left-handed or right-handed? Left-footed or right-footed? Is your right eye dominant or is it your left?

Around 90% of the world's population is right-handed. Why most people favour the right side is not completely understood by scientists. Some think that the reason is related to which side of your brain you use for language. The right side of your body is controlled by the left side of your brain, and in around 90% of people the left side of the brain also controls the language.

Others think the reason might have more to do with culture. The word, 'right' is associated with being correct and doing the right thing, while the word, 'left' originally meant, 'weak'. Favouring the right hand may have become a social development as more children were taught important skills by right-handed people and various tools were designed to be used with the right hand.

Around 80% of the people are right-footed and about 70% favour their right eye. However, these percentages are lower than those who are right-handed and this could be because your body has more freedom of choice in choosing its favoured foot and eye than that of its favoured hand. In other words, you are more likely to be trained to use your right hand than your right foot and even more so than your right eye.

It's not strange to find people who favour the opposite hand and foot (e.g. left hand and right foot), and some people are lucky enough to be ambidextrous, meaning they can use their left and right sides with equal skill.

Try testing others and come to your own conclusions about which side the human body generally favours and why.

What's more

Think & find out: Are you more likely to be left-handed if one of your parents is left-handed? What are some of the possible disadvantages for left-handed people? (Tools, writing materials, etc.) Do left-handed people have an advantage in sports?

Interesting fact: In 2009, only 7% of the players in the National Basketball Association (NBA) were left handed while in 2008, around 26% of the Major League Basketball (MLB) pitchers were left-handed.

Is it better to be left-handed in some sports than others? What do you think?

Design & Test a Parachute

Learn about air resistance while making an awesome parachute! Design one that can fall slowly to the ground before putting it to the test, making modifications as you go.

What you need

A plastic bag or light material, Scissors, String, A small object to act as the weight or a little action figure would be perfect.

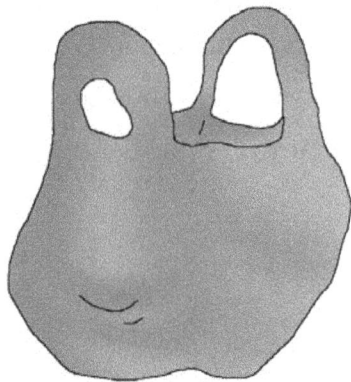

What to do

1. Cut out a large square from your plastic bag or material.
2. Trim the edges so that it looks like an octagon. (an eight-sided shape).
3. Cut a small whole near the edge of each side.
4. Attach eight pieces of string of the same length to each of the holes.

5. Tie the pieces of string to the object you are using as a weight.
6. Use a chair or find a high spot to drop your parachute and test how well it works. Remember that you want it to drop as slow as possible.

What's happening

Hopefully your parachute will descend slowly to the ground, giving your weight a comfortable landing. When you release the parachute, the weight pulls down on the strings and opens up a large surface area of material that uses air resistance to slow it down. The larger the surface area, the more the air resistance and the slower the parachute will drop.

Cutting a small hole in the middle of the parachute will allow the air to slowly pass through it rather than spilling out over on one side. This should help the parachute fall straighter.

Make a Big Dry Ice Bubble

Have fun making a big dry ice bubble that will grow and grow as it fills with fog. This experiment is a great one for adults as well as kids. Add water to the dry ice, cover it with a layer of soapy water and watch your bubble grow. How big will it get before it bursts? Give it a try and find out!

What you need

Water, A large bowl with a lid around the top (a smaller bowl or cup will work too), A strip of material or cloth, Soapy mixture for making bubbles (water and some dishwashing liquid should do the trick), Dry ice – one piece for a cup, more for a bowl. Places where adults can buy dry ice such as ice cream stores, grocery shops, etc.

Note:

Be careful with dry ice as it can cause skin damage if not used safely. Adults should handle dry ice with gloves and avoid directly breathing in the vapour.

What to do

1.   Place your dry ice in the bowl and add some water. (It should start looking like a spooky cauldron).
2.   Soak the material in your soapy mixture and run it around the lid of the bowl before dragging it across the top of the bowl to form a bubble layer over the dry ice.
3.   Stand back and watch your bubble grow!

What's happening

Dry ice is carbon dioxide (CO_2) in its solid form. At temperatures above -56.4 °C (-69.5 °F), dry ice changes directly from a solid to a gas, without ever being a liquid. This process is called *sublimation*. When dry ice is put in water, it accelerates the sublimation process, creating clouds of fog that fill up your dry ice bubble until the pressure becomes too much and the bubble explodes, spilling fog over the edge of the bowl. Dry ice is sometimes used as a part of theatre productions and performances to create a dense foggy effect. It is also used to preserve food, freeze lab samples and even to make ice cream!

Diet Coke & Mentos Eruption

Oone of the most popular experiments of modern times is the Diet Coke and Mentos Geyser. Made popular by Steve Spangler, this experiment is a lot of fun and sure to amaze your friends and family. (Assuming, you do it outside rather than in the living room).

What you need

A large bottle of Diet Coke, About half a pack of Mentos and Geyser tube (optional, but makes things much easier).

What to do

1. Make sure you are doing this experiment in a place where you won't find any difficulty in getting a Diet Coke. Outside on some grass or open area is perfect. Please don't try this one in your family lounge!!
2. Stand the Diet Coke upright and unscrew the lid. Put some sort of funnel or tube on top of it so that you can drop the Mentos in it at the same time. (About half the pack is a good amount). Doing this part can be tricky if you don't have a specially designed geyser tube. Buying one from a local store is recommended.

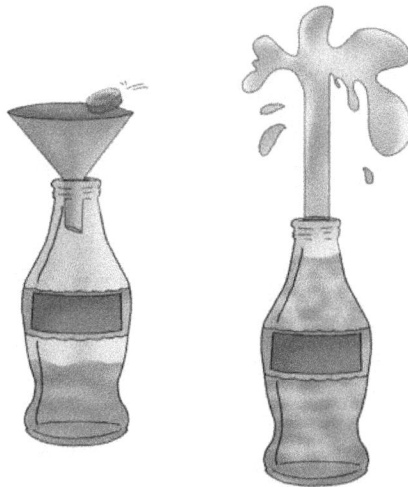

3. Time for the fun part. Drop the Mentos into the Diet Coke and run like a mad person! If you've done it properly, a huge geyser of Diet Coke should come flying out of the bottle. It's a very impressive sight. The record is about 9 metres (29 feet) high!

What's happening

Although there are a few different theories about how this experiment works, the most favoured and acceptable reason is because of the combination of carbon dioxide in the Diet Coke and the little dimples found on Mentos candy pieces.

The thing that makes soda drinks bubbly is the carbon dioxide that is pumped in when they bottle the drink at the factory. It doesn't get released from the liquid until you pour it into a glass and drink it. Some gas also is released when you open the lid (more if you shake it up beforehand). This means that there is a whole lot of carbon dioxide gas just waiting to escape the liquid in the form of bubbles.

Dropping something into the Diet Coke speeds up this process by both breaking the surface tension of the liquid and also allowing the bubbles to form on the surface area of the Mentos. Mentos candy pieces are covered in tiny dimples (a bit like a golf ball), which dramatically increases the surface area and allows a huge amount of bubbles to form.

The experiment works better with Diet Coke than other sodas due to its slightly different ingredients and the fact that it isn't so sticky. It has been found that Diet Coke that had been bottled more recently worked better than older bottles that might have lost some of their fizz lying on shop shelves for too long. So, just check the bottle for the date, of packaging to get the right fizz!

Blowing Up Balloons with CO2

Chemical reactions make for some great experiments. Make use of the carbon dioxide given off by a baking soda and lemon juice reaction by funnelling the gas through a soft drink bottle. Blowing up balloons was never so easy!

What you need

A balloon, About 40 ml of water (a cup is about 250 ml so you don't need much), A soft drink bottle, A drinking straw, Juice from a lemon and 1 teaspoon of baking soda.

What to do

1. Before you begin, make sure that you stretch out the balloon to make it as easy as possible to inflate.
2. Pour the 40 ml of water into the soft drink bottle.
3. Add the teaspoon of baking soda and stir it around with the straw until it has dissolved.
4. Pour the lemon juice in and quickly put the stretched balloon over the mouth of the bottle.

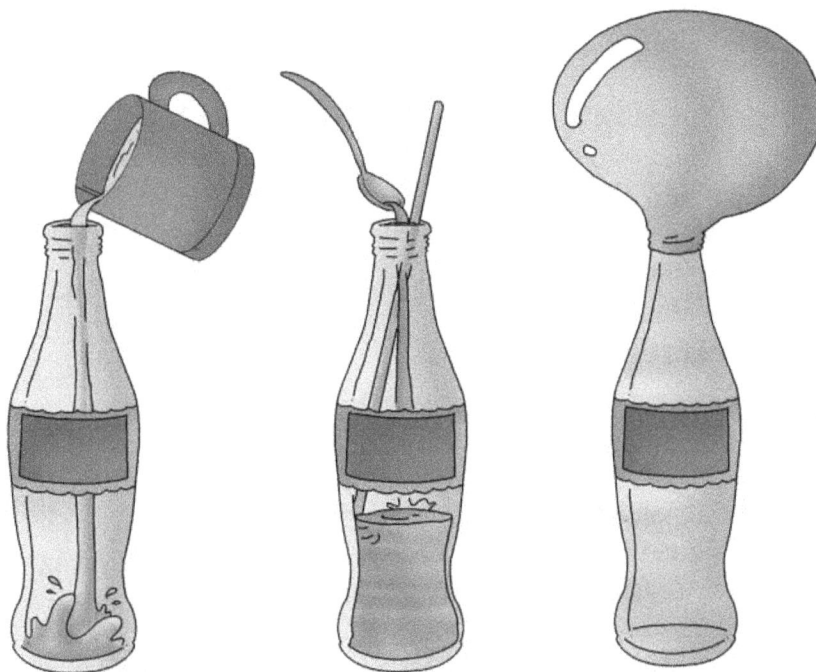

What's happening

If all goes well then your balloon should inflate! Adding the lemon juice to the baking soda creates a chemical reaction. The baking soda is a base, while the lemon juice is an acid. When the two combine, they create carbon dioxide (CO_2). The gas rises up and escapes through the soft drink bottle, it doesn't however escape the balloon, pushing it outwards and blowing it up. If you don't have any lemons then you can substitute the lemon juice for vinegar.

Make your Own Fake Snot

As disgusting as it might sound to some people, let's make some fake snot! Snot actually serves an important purpose in our body. So this experiment is not all about grossing out our friends, although that's certainly a part of the fun.

What you need

Boiling water (be careful with this), A cup, Gelatin, Corn syrup, A teaspoon and A fork

What to do

1. Fill half a cup with boiling water.
2. Add three teaspoons of gelatin to the boiling water.
3. Let it soften before stirring with a fork.
4. Add a quarter of a cup of corn syrup.
5. Stir the mixture again with your fork and look at the long strands of gunk that have formed.
6. As the mixture cools slowly, add more water, small amounts at a time.

What's happening

Mucus is made mostly of sugars and protein. Although different than the ones found in the real thing, this is exactly what you use to make your fake snot. The long, fine strings you could see inside your fake snot when you moved it around are protein strands. These protein strands make the snot sticky and capable of stretching.

Make a Tornado in a Bottle

Learn how to make a tornado in a bottle with this incredible science experiment for kids. Using easy-to-find items, such as dish washing liquid, water, glitter and a bottle, you can make your own *mini tornado* and that's a lot safer than the one you might see on the weather channel. Follow the instructions and enjoy the cool water vortex you create!

What you need

Water, A clear plastic bottle with a cap (that won't leak), Glitter and Dish washing liquid.

What to do

1. Fill the plastic bottle with water until it reaches around three quarters full.
2. Add a few drops of dish washing liquid.
3. Sprinkle in a few pinches of glitter. (This will make your tornado easier to see).
4. Put the cap on tightly.
5. Turn the bottle upside down and hold it by the neck. Quickly spin the bottle in a circular motion for a few seconds. Stop and look inside to see if you can see a *mini tornado* forming in the water. You might need to try it a few times before you get it working properly.

What's happening

Spinning the bottle in a circular motion creates a *water vortex* that looks like a *mini tornado*. The water is rapidly spinning around the centre of the vortex due to centripetal force (an inward force directing an object or fluid such as water towards the centre of its circular path). Vortexes found in nature include tornadoes, hurricanes and waterspouts. (A tornado that forms over water).

Cut Ice Cubes in Half like Magic

Speed up the melting process of ice with the help of a little pressure. Cut a piece of ice in half like magic and learn how this process relates to ice skating.

What you need

One ice cube, A piece of fishing line (or something similar) with a weight tied to each end, A container and some kind of a tray to keep things from getting wet.

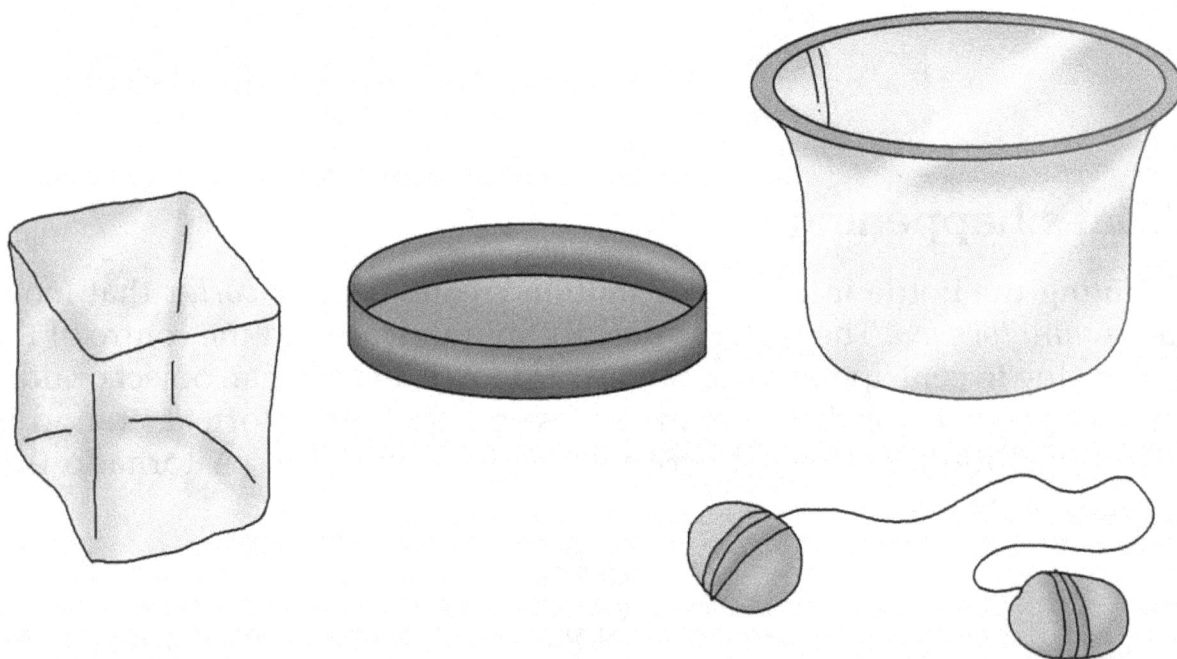

What to do

1. Turn the container upside down and put it on the tray.
2. Place the ice cube on top of the upside down container.
3. Rest the fishing line over the ice cube so that the weights are left dangling over the side of the container.
4. Watch it for around 5 minutes.

What's happening

The pressure from the two weights pulls the string through the ice cube by melting the ice directly under the fishing line. This is similar to ice skating where the blades of a skater melt the ice directly underneath, allowing the skater to move smoothly on a thin layer of water.

Static Electricity Experiment

Find out about positively and negatively charged particles using a few basic items. Can you control if they will be attracted or repelled each other?

What you need

2 inflated balloons with string attached, Your hair, Aluminium can, and Woollen fabric.

What to do

1. Rub the two balloons one by one against the woollen fabric. Then try moving the balloons together. Do they want to attract, or are they repelled each other?

2. Rub one of the balloons back and forth on your hair and then slowly pull it away. Ask someone nearby what they can see or if there's nobody else around try looking in a mirror.

3. Put the aluminium can on its side on a table, after rubbing the balloon on your hair again. Hold the balloon close to the can and watch as it rolls towards it. Slowly move the balloon away from the can and it will follow.

What's happening

Rubbing the balloons against the woollen fabric or your hair creates *static electricity*. This involves negatively charged particles (electrons) jumping to positively charged objects. When you rub the balloons against your hair or the fabric, they become negatively charged because they have taken some of the electrons from the hair/fabric and left them positively charged.

They say opposites attract and that is certainly the case in these experiments. Your positively charged hair is attracted to the negatively charged balloon and starts to rise up to meet it. This is similar to the aluminium can which is drawn to the negatively charged balloon as the area near it becomes positively charged. Once again, opposites attract.

In the first experiment, both the balloons were negatively charged after rubbing them against the woollen fabric and so, they repelled each other.

What Absorbs More Heat?

When you're out in the sun on a hot summer day, it is convenient to wear some light coloured clothes, but why is that? Experiment with light, colour, heat and some water to find out.

What you need

2 identical drinking glasses or jars, Water, A thermometer, 2 elastic bands or some cellotape, White paper and a Black paper.

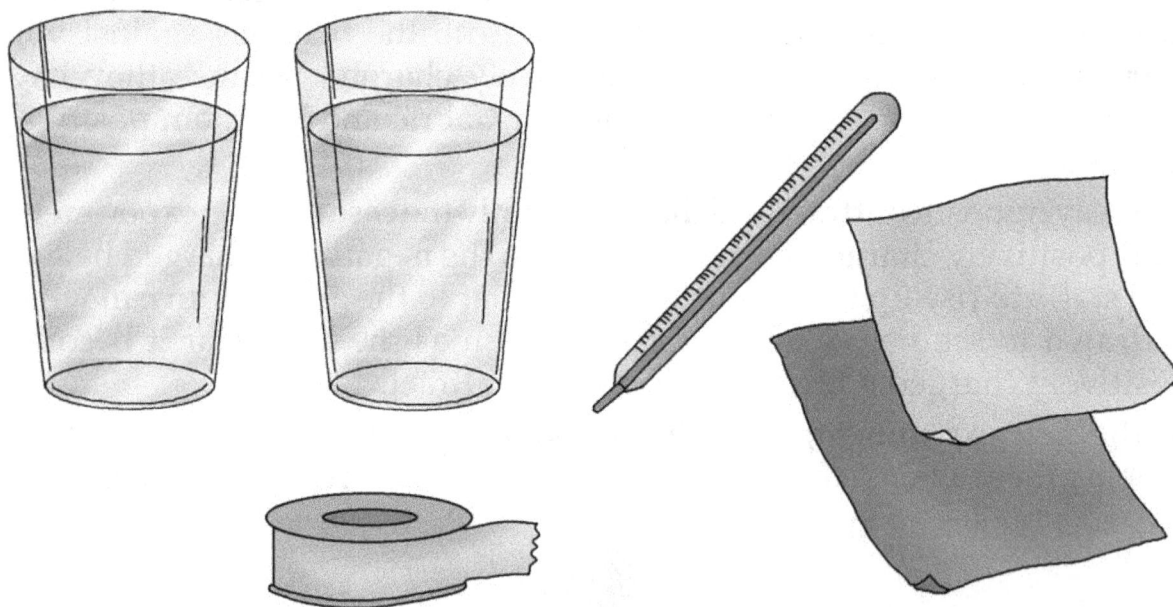

What to do

1. Wrap the white paper around one of the glasses using an elastic band or cellotape to hold it on.
2. Do the same with the black paper and the other glass.
3. Fill the glasses with same amount of water.
4. Leave the glasses out in the sun for a couple of hours before returning to measure the temperature of the water in each glass.

What's happening

Dark surfaces, such as the black paper absorb more light and heat than the lighter ones, such as the white paper. Now measure the temperatures of the water. The glass with the black paper around it should be hotter than the other. Lighter surfaces reflect more light, and that's why people wear light coloured clothes in summer as it keeps them cooler.

Water Molecules on the Move

This experiment is great for testing if hot water molecules really move faster than the cold ones. Pour some water, drop in some food colouring and compare the results.

What you need

A clear glass filled with hot water, A clear glass filled with cold water, Food colouring and an eye dropper.

What to do

1. Fill the glasses with the same amount of water, one cold and one hot.
2. Put one drop of food colouring into both the glasses as quickly as possible.
3. Watch what happens to the food colouring.

What's happening

If you watch closely, you will notice that the food colouring spreads faster throughout the hot water than in the cold. The molecules in the hot water move at a faster rate, spreading the food colouring faster than the cold water molecules which move slower.

Plant Seeds & Watch them Grow

Learn about seed germination with this fun science experiment for kids. Plant some seeds and follow the growth of the seedlings as they sprout from the soil while making sure to take proper care of them with just the right amount of light, heat and water.

What you need

Select fresh seeds of your choice, such as pumpkin seeds, sunflower seeds, lima beans or pinto beans, Good quality soil (loose, aerated, with lots of peat moss), if you don't have any you can buy some potting soil at your local garden store, A container to hold the soil and your seeds, Water, Light and Heat.

What to do

1. Fill the container with soil.
2. Plant the seeds inside the soil.
3. Place the container somewhere warm. Sunlight is good but try to avoid too much direct sunlight. A window sill is a good spot.

4. Keep the soil moist by watering it everyday (be careful not to use too much of water).
5. Record your observations as the seeds germinate and the seedlings begin to sprout from the seeds.

What's happening

Hopefully after a week of looking after them, your seedlings will be on their way. Germination is the process of a plant emerging from a seed and beginning to grow. For seedlings to grow properly from a seed, they need the right conditions. **Water** and **oxygen** are required for seeds to germinate. Many seeds germinate at a temperature just above the normal room temperature but others respond better to warmer temperatures, cooler temperatures or even changes in temperature. While light can be an important factor for germination, some seeds actually need darkness to germinate. If you buy seeds, read the specific instructions written on its packet for its proper germination.

Continue to look after your seedlings and monitor their growth. For further experiments, you could compare the growth rates of different types of seeds or the effect of different conditions on their growth.

Taste Testing without Smell

We all know that some foods taste better than others, but what gives us the ability to experience all these unique flavours? This simple experiment shows that there's a lot more to taste than you might have first thought.

What you need

A small piece of peeled potato, A small piece of peeled apple (same shape as that of the potato so that you can't tell the difference).

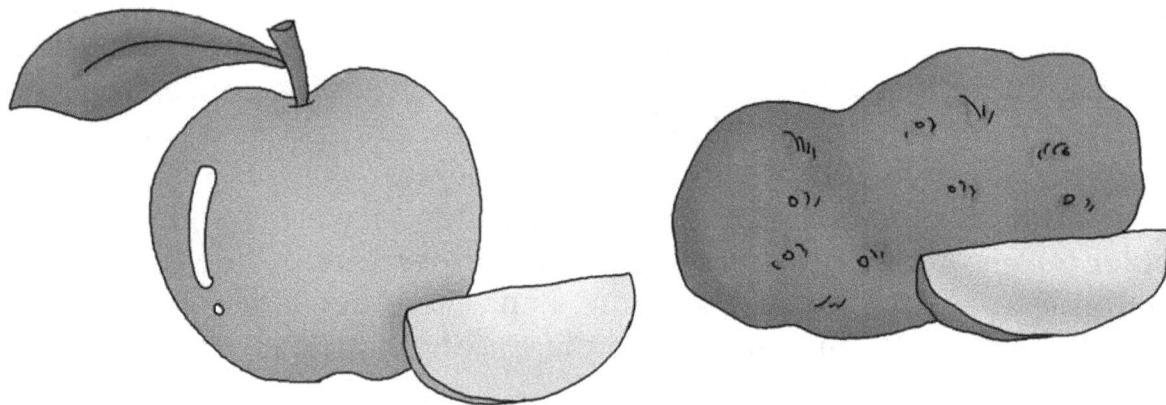

What to do

1. Close your eyes and mix up the piece of potato and the piece of apple so that you don't know which is which.
2. Hold your nose and eat each piece, can you tell the difference?

What's happening

Holding your nose while tasting the potato and the apple makes it hard to tell the difference between the two. Your nose and mouth are connected through the same airway which means that you taste and smell foods at the same time. Your sense of taste can recognise salty, sweet, bitter and sour but when you combine this with your sense of smell you can recognise many other individual 'tastes'. Take away your smell, (and sight) and you limit your brain's ability to tell the difference between certain foods.

Escaping Water

Water can certainly move in mysterious ways. Get the water from one cup to make its way up hill and back down into a second empty cup with the help of paper towels and an interesting scientific process.

What you need

A glass of water, An empty glass and Some paper towels.

What to do

1. Twist a couple of pieces of paper towel together until it forms something that looks a little like a piece of rope. This will be the 'wick' that will absorb and transfer the water (a bit like the wick on a candle transferring the wax to the flame).
2. Place one end of the paper towels into the glass filled with water and the other into the empty glass.
3. Watch what happens. (This experiment requires a little bit of patience).

What's happening

Your paper towel rope (or wick) starts getting wet. After a few minutes you will notice that the empty glass is getting filled with water. It keeps filling until there is an even amount of water in each glass. How does this happen?

This process is called the 'capillary action'. Water uses this process to move along the tiny gaps in the fibre of the paper towels. It occurs due to the adhesive force between the water and the paper towel being stronger than the cohesive forces inside the water itself. This process can also be seen in plants where moisture travels from the roots to the rest of the plant.

Microscopic Creatures in Water

Water can be home to a lot of interesting creatures and microorganisms, especially if it's dirty water found in ponds or near plants. Take some samples, view them under a microscope and see what you can find. How clean is the water from your tap compared to the water found in a pond? Experiment and see what kind of microscopic creatures you can find!

What you need

A concave slide, A dropper, A microscope, Different samples of water (tap water, pond water, muddy water, etc). Water near plants or in the mud are good places to take samples as they usually contain more microorganisms.

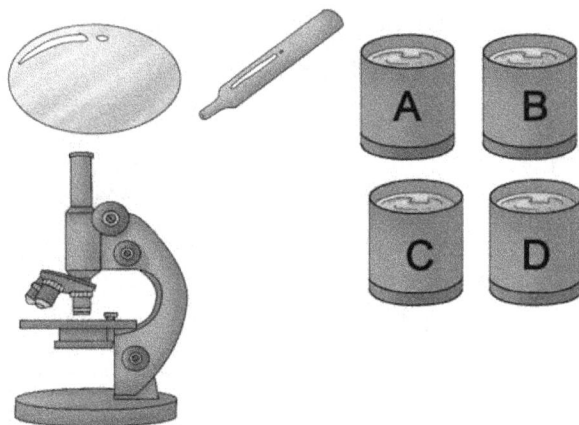

What to do

1. Set up your microscope, preferably using its highest setting.
2. Use the dropper to take some water from one of your samples.

Now put it on the concave slide. Focus the microscope, what can you see? Be patient if you can't see anything. If you still can't see anything and have checked that you are in focus, try a different water sample.

3. Look at how the creatures move. After observing their movements, you might like to record their behaviour and draw them.

What are you looking at?

Some of the creatures and microorganisms you might be able to see include:

> **Euglenas -** They have a long tail called a flagellum which allows them to move.

> **Protozoa -** They have a flagella (tail) which can be hard to see. The difference between a protozoa and an algae is often hard to define.

> **Amoebas -** These microorganisms swim by wobbling. They also surround their foods like a blob in order to eat it.

> **Algae -** Not considered to be plants by most scientists, these organisms might be coloured yellowish, greenish or reddish. They may also be found by themselves or in chains.

There might even be larger creatures, such as worms or brine shrimps in your water samples, depending on where you took them from.

Bend a Straw with your Eyes

Using the power of your eyes, bend a straw placed in half a glass of water without even touching it! It sounds like magic but it's really another amazing scientific principle at work.

What you need

A glass half filled with water, A straw, 2 eyes (preferably yours).

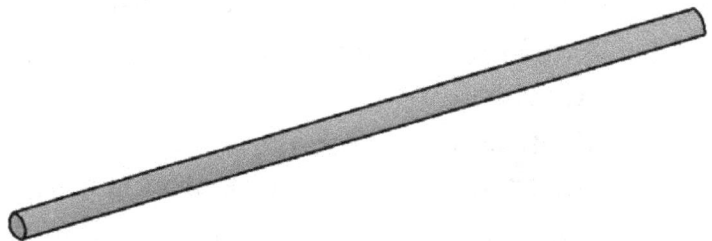

What to do

1. Look at the straw from the top and bottom of the glass.
2. Look at the straw from the side of the glass. Focus on the point where the straw enters the water. What is strange about what you see?

What's happening

Our eyes are using light to see various objects all the time, but when this light travels through different mediums (such as water and air), it changes direction slightly. Light refracts (or bends) when it passes from water to air. The straw looks bent because you are seeing the bottom part through the water and air but the top part through the air only. Air has a refractive index of around 1.0003, while water has a refractive index of about 1.33.

Make your Own Rainbow

Learn how to make a rainbow with this fun science experiment. Using just a few simple everyday items, you can find out how rainbows work while enjoying an interactive and interesting activity that's perfect for kids.

What you need

A glass of water (about three quarters full), White paper and A sunny day.

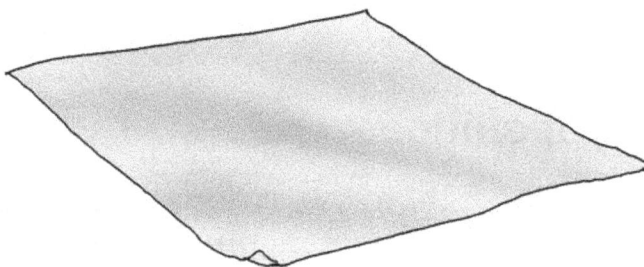

What to do

1. Take the glass of water and paper to a part of the room with sunlight (near a window is good).
2. Hold the glass of water (being careful not to spill it) above the paper and watch as sunlight passes through the glass of water.

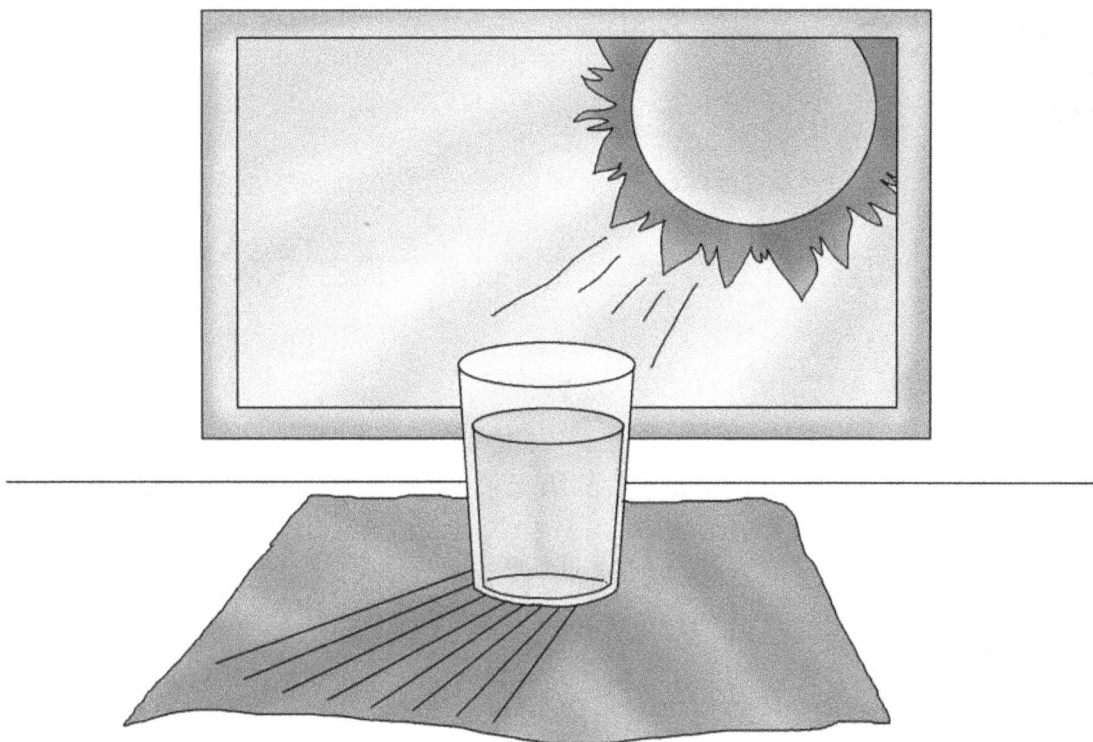

The sunlight refracts (bends) and forms a rainbow of colours on your sheet of paper.

3. Try holding the glass of water at different heights and angles to see if it has a different effect.

What's happening

While you normally see a rainbow as an arc of colour in the sky, they can also form in other situations. You may have seen a rainbow in a water fountain or in the mist of a waterfall and you can even make your own such as you did in this experiment.

Rainbows form in the sky when sunlight refracts (bends) as it passes through the raindrops. It acts in the same way when it passes through your glass of water. The sunlight refracts, separating it into the colours violet, indigo, blue, green, yellow, orange and red (VIBGYOR).

Warm Air needs more Room

As its temperature rises, air starts to act a little differently. Find out what happens to a balloon when the air inside it heats up.

What you need

An empty bottle, A balloon and A pot of hot water (not boiling).

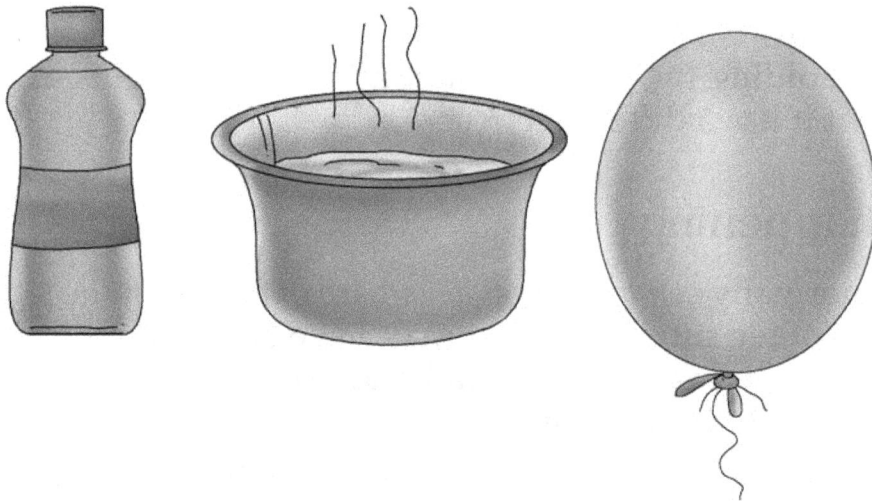

What to do

1. Stretch the balloon over the mouth of the empty bottle.
2. Put the bottle in the pot of hot water, and let it stand in the position for a few minutes. Watch what happens.

What's happening

As the air inside the balloon heats up, it starts to expand. The molecules begin to move faster and further apart from each other. This is what makes the balloon stretch. There is still the same amount of air inside the balloon and the bottle, it has just expanded as it heats up.

Warm air therefore takes up more space than the same amount of cold air. It also weighs less than cold air occupying the same space. You might have seen this principle in action if you have flown in or watched a hot air balloon. Amazing, isn't it?

Bending Water
with Static Electricity

There is an easy and fun science experiment that is great for helping kids learn about static electricity.

Try bending water with static electricity produced by combing your hair or rubbing it with an inflated balloon. Can it really be done? Give it a try and find out!

What you need

A plastic comb (or an inflated balloon), A narrow stream of water from a tap and Dry hair.

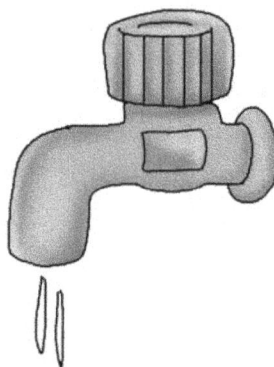

What to do

1. Turn on the water so that it falls from the tap in a narrow stream (just a few millimeters across but not in droplets).
2. Run the comb through your hair just as you normally would when brushing it. Do this around 10 times). If you are using a balloon, then rub it back and forth against your hair for a few seconds.

3. Slowly move the comb or balloon towards the stream of water (without touching it) while watching closely to see what happens.

What's happening

The static electricity you built up by combing your hair or rubbing it against the balloon attracts the stream of water, bending it towards the comb or balloon like magic!

Negatively charged particles called electrons jump from your hair to the comb as they rub together. The comb now has extra electrons and is negatively charged. The water features both positive and negatively charged particles and is neutral. Positive and negative charges are attracted to each other. So when you move the negatively charged comb (or balloon) towards the stream, it attracts the water's positively charged particles and the stream bends!

Steel Wool & Vinegar Reaction

Soak steel wool in vinegar and watch what happens as the iron in the steel begins to react with the oxygen around it. This incredible science experiment for kids is great for learning about chemical reactions.

What you need

Steel wool, Vinegar, Two beakers, Paper or a lid (something to cover the beaker to keep the heat in), Thermometer.

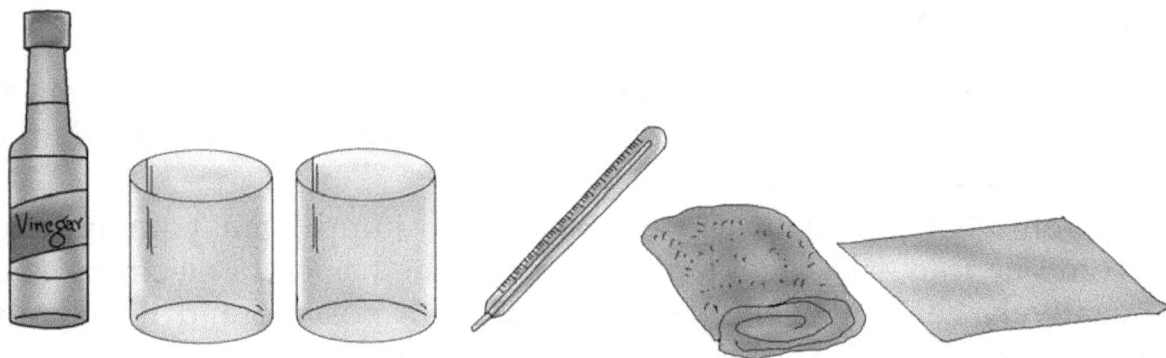

What to do

1. Place the steel wool in a beaker.
2. Pour vinegar on to the steel wool and allow it to soak in the vinegar for around one minute.
3. Remove the steel wool and drain any excess vinegar.

4. Wrap the steel wool around the base of the thermometer and place them both in the second beaker.

5. Cover the beaker with paper or a lid to keep the heat in. (Make sure you can still read the temperature on the thermometer. Having a small hole in the paper or lid for the thermometer to go through is a good idea).

6. Check the initial temperature and then monitor it for around five minutes.

What's happening

The temperature inside the beaker should gradually rise. You might even notice the beaker getting foggy. When you soak the steel wool in vinegar, it removes the protective coating of the steel wool and allows the iron in the steel to rust. *Rusting (or oxidation)* is a chemical reaction between iron and oxygen. This chemical reaction creates heat energy which increases the temperature inside the beaker. This experiment is an example of an *exothermic reaction,* a chemical reaction that releases energy in the form of heat.

Energy Transfer through Balls

Energy is constantly changing forms and transferring between objects. Try seeing for yourself how this works. Use two balls to transfer kinetic energy from the big ball to the smaller one and see what happens.

What you need

A large, heavy ball, such as a basketball or a soccer ball, A smaller, light ball, such as a tennis ball or an inflatable rubber ball.

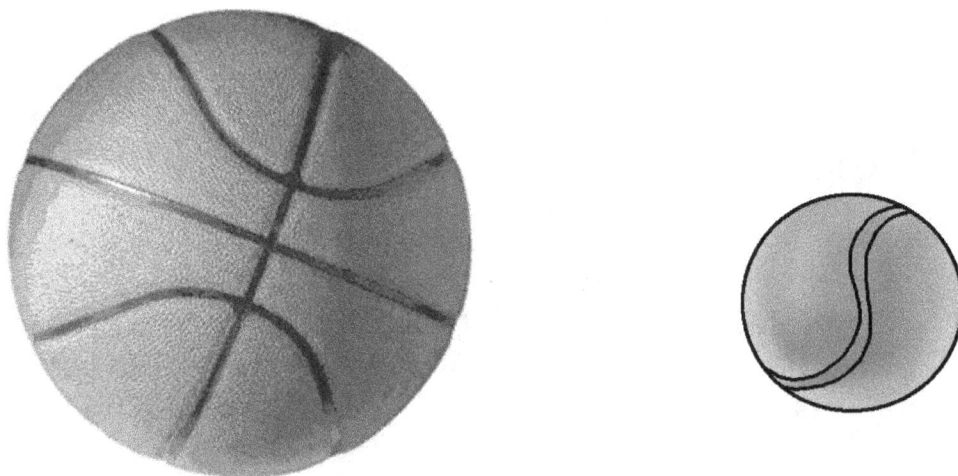

What to do

1. Make sure you are outside with plenty of space.
2. Carefully put the tennis ball on top of the basketball, holding one hand under the basketball and the other on top of the tennis ball.

3. Let go of both the balls at exactly the same time and observe what happens.

What's happening

If you dropped the balls at the same time, the tennis ball should bounce off the basketball and fly high into the air. The two balls hit each other just after they hit the ground. A lot of the kinetic energy in the larger basketball is transferred through to the smaller tennis ball, sending it high into the air.

While you held the balls in the air before dropping them, they had another type of energy called the 'potential energy'. The balls gained this energy through the effort it took you to lift the balls up. It is interesting to note that energy is never lost, only transferred into other kinds of energy.

Make a String Telephone

Step back in time and use some old fashioned technology to make a string phone while learning about sound waves with this incredible science project.

What you need

2 paper cups, A sharp pencil or sewing needle to help poke holes, A string (kite string and fishing lines work well).

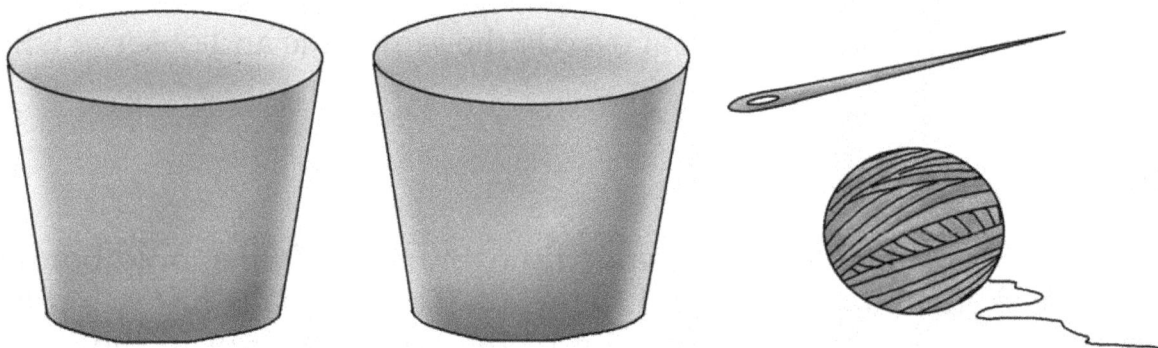

What to do

1. Cut a long piece of the string. You can experiment with different lengths but perhaps 20 metres (66 feet) is a good length to start with.
2. Poke a small hole in the bottom of each cup.
3. Thread the string through each cup and tie knots at each end to stop it pulling through the cup. (Alternatively, you can use a paper clip, washer or similar small object to hold the string in place).
4. Move into a position with you and your friend holding the cups at a distance that makes the string tight (making sure the string isn't touching anything else).
5. One person talks into the cup while the other puts the cup to his/her ear and listens. Can they hear each other?

What's happening

Speaking into the cup creates sound waves which are converted into vibrations at the bottom of the cup. The vibrations travel along the string and are converted back into sound waves at the other end. So your friend can hear what you said. Sound travels through the air but it travels even better through solids, such as your cup and string, allowing you to hear sounds that might be too far away when travelling through the air.

What's more

Landline telephones feature microphones that convert *sound waves into electric currents* that are then sent through wires and converted back into sound waves by an earphone inside the telephone at the other end. Modern mobile phones use radio waves (part of the electromagnetic spectrum that includes microwaves, infrared, visible light, X-rays and others) to communicate with base stations located throughout telephone networks.

Phones have come a long way since Alexander Graham Bell was awarded the first electric telephone patent by the United States Patent and Trademark Office, back in 1876. Today's cellphones are a marvel of modern technology, featuring not only the ability to make phone calls but to also surf the web, play music, view documents and much more.

Can you Protect a Falling Egg?

The egg drop is a classic science project that kids will love. Can you design a system that will protect an egg from a fall? Give it a try and find out.

Use items from around the house to build something that will prevent eggs smashing all over the ground.

What you need

Eggs, Paper towels.

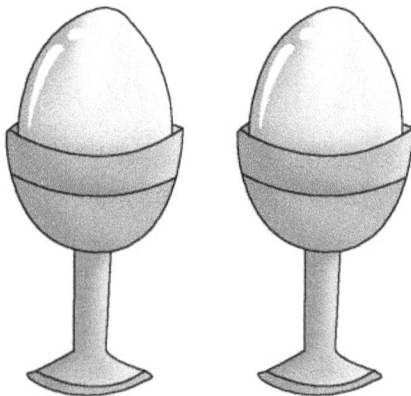

Note:

Plastic straws, Popsicle sticks, Tape, Recycled paper, Glue, Plastic bags, Boxes, Used material and Plastic containers.

What's your aim

➢ Your goal is simple; design and build a system that will protect an egg from a 1 metre (3.3 feet) drop. Eggs that smash or crack fail the test while eggs that survive without a scratch pass!

What to do

You need to create something that can absorb the energy the egg gathers as it accelerates towards the ground. A hard surface will crack the egg so you have to think carefully about how you can protect it. Something that will cushion the egg at the end of its fall is a good place to start. You want the egg to decelerate slowly so that it doesn't crack or smash all over the ground. You'll need to undergo a few trials. So have some eggs ready as guinea pigs. Those that don't survive will at least be comforted knowing they were smashed for a good cause, and if not, you can at least have scrambled eggs for dinner, isn't it?

Grow your Own Salt Crystals

Have fun growing your own salt crystals with this simple experiments. You can do further research with a microscope once you're finished. Crystals are beautiful to look at and you might even want to start your own collection.

What you need

A jar, Water, About half a cup of salt, A spoon for stirring, A string, Scissors amd 2 toothpicks.

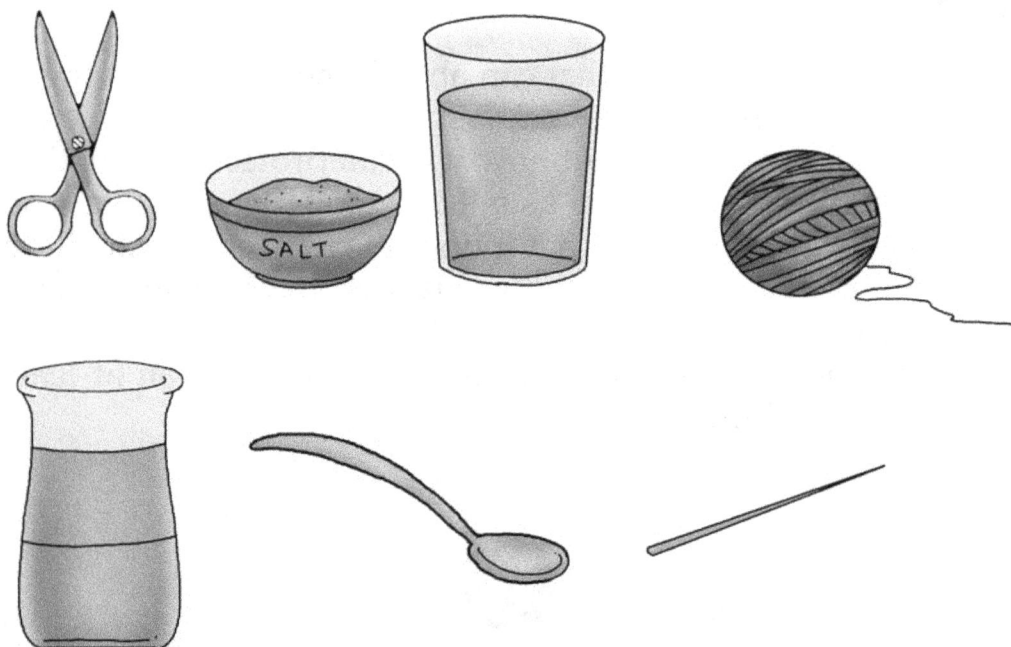

What to do

1. Fill the jar with water.
2. Add about half a cup of salt to the water.
3. Mix the solution together with a spoon.
4. Cut a piece of the string with scissors and tie each end to a toothpick.
5. Place the string over the top of the jar so that the string dangles into the middle of the solution and the toothpicks hang over the edge.
6. Don't forget to clean up when you have finished.

What's next?

Leave the experiment and wait for salt crystals to form along the string. They are an excellent example of cubic crystals and you can do further research with them by examining them under a microscope.

When you look at various crystals under a microscope, you can examine the differences between them: Are they perfectly formed? What shape are they? What colour? Can you see any microorganism on the crystals?

Crystals can be found grouped together as lots of small crystals or as huge individual crystals. They vary in size from those at the microscopic level to crystals that are metres in length!

Try collecting a range of crystals for your project, label the different types and make a rock collection box to keep them in.

Make your Own Robot!

Get creative and make a robot! These fun projects for kids are aimed at students of different levels. Younger children can enjoy using a range of household items as they build robots with features that are only limited to what their imagination can come up with.

Build a robot from household items

Let kids enjoy building a robot from everyday household items. It's lots of fun and is sure to keep their attention.

What you need

Useful materials include soft drink lids, Old boxes, Tin foil, Ice cream containers, Old clothing, Various materials such as straws, paper and crayons.

What to do

You'll need quite a lot of materials (depending on how many children will be taking part). A good idea is to start off with unused cardboard boxes and proceed from there. The children can use glue or tape boxes together to form the general shape of a robot before attaching other items to complete the project. There is room for a wide variety of ideas on this project. So if you have an idea that you think will work, then give it a go!

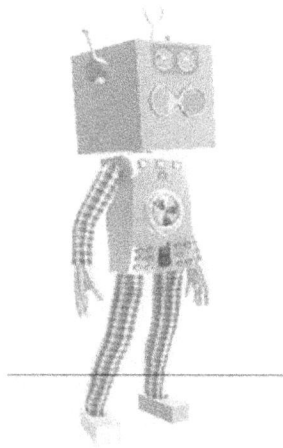

Build a robot using electronic equipment or a robotics kit set

For the older groups of children, you can try a robot building project using real electronic equipment or a robotics kit set.

What you need

There are a number of great robotics kit sets such as the dependable Lego Mindstorms NXT, which offers plenty of scope for robot building challenges.

What to do

Rather than just letting them build any type of robot, give them a fun challenge which can serve as an inspiration behind the design of their robot as well as the focus of any program they make using a computer. This challenge could involve a race of some type, robots that use sensors to find something, or a test of strength or building a robot that responds to some form of human input. During designing and building their robot, students will have to think about how they will program it as well. This project can be further developed into a great science project focusing on technology. You could research what kind of artificial intelligence your robot is capable of as well as any physical limitations it has that stop it from performing the required tasks.

Make your Own Fossil

Find an interesting object and set it in stone, letting its impression live on in the form of a fossil.

Have fun making your own fossil and learning how scientists use them to unlock secrets of the past, including those that provide a remarkable insight into life in the age of the dinosaurs.

What you need

Plasticine, 2 paper cups, An object that you would like to use as the fossilized impression, Plaster of Paris and Water.

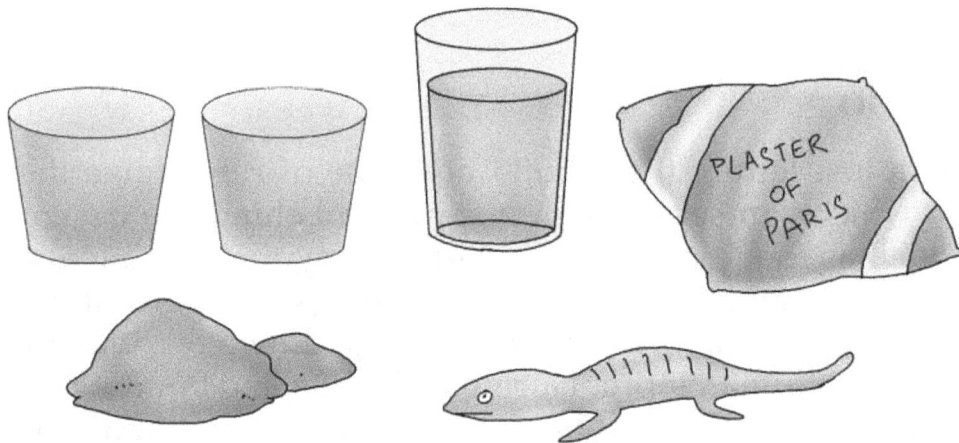

What to do

1. Flatten a ball of plasticine until it is about 2 cm thick while making sure the top is smooth.
2. Put the plasticine inside a paper cup with the smooth side facing up. Carefully press the object you want to fossilize into the plasticine until it is partially buried.

3. Carefully remove the object from the plasticine. An impression of the object should be left behind.

4. Pour half a cup of plaster of Paris into the other paper cup. Add a quarter cup of water to the plaster and stir until the mixture is smooth. Leave it for around two minutes.

5. When the mixture has thickened, pour it on top of the plasticine in the other cup. Leave the mixture until the plaster has dried.

6. When the plaster has fully dried, tear away the sides of the paper cup and take out the plasticine and the plaster. Keep it in a warm dry place and enjoy your very own fossil.

What's happening

Fossils are extremely useful records of the past. In your case, you left behind an impression of an object you own but fossils found by scientists around the world can date back to the time of dinosaurs. These fossils allow the palaeontologists (scientists who study these types of fossils) to study what life might have been about millions of years ago. Fossils, such as the one you made can leave delicate patterns and a surprising amount of detail.

Make Stalactites & Stalagmites

Stalactites and stalagmites found in caves are an impressive feature of nature, but they take a long time to form. Let's speed up the process by making our own stalactites and stalagmites.

With enough time, they might even join in the middle and form a column.

What you need

Two glass jars, A saucer, Woollen thread, Either baking soda, Washing soda or Epsom salts.

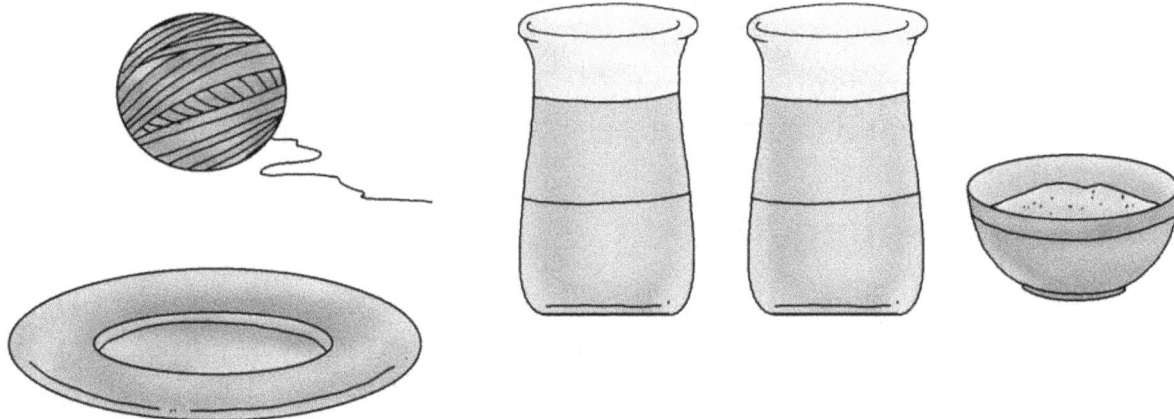

What to do

1. Fill both the jars with hot water. Dissolve as much soda as you can into each one.
2. Place the two jars in a warm place and put the saucer between them.

3. Twist several strands of woollen thread together before dipping the ends into the jars and letting the middle of the thread hang down above the saucer. The ends can be weighed down with various small, heavy objects to keep them in the jars.
4. The two solutions should creep along the thread until they reach the middle and then drip down onto the saucer.
5. Watch what happens to the experiment over the next few days.
6. Don't forget to wash your hands when you've finished.

What's happening

Over a few days, the dripping water will leave behind the baking soda, forming a tiny stalactite (which forms from the roof) and stalagmite (which forms from the ground). With enough time, these may eventually join to create a single column. Stalactites and stalagmites are columns of stone which form in underground caves. They are made from minerals dissolved in rainwater that slowly drips from the roofs and walls of caves.

Make your Own
Kaleidoscope

Enjoy the beautiful colours and symmetrical patterns formed by a kaleidoscope by making your own one.

Understand how light bounces between the mirrors of your kaleidoscope and have fun decorating it when finished.

What you need

3 pieces of mirrored perspex, A roll of duct tape or masking tape, Overhead transparency paper, Coloured see-through plastic, A pencil.

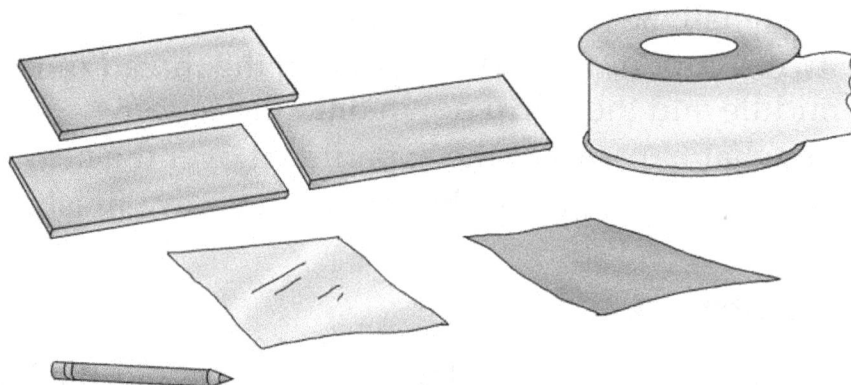

What to do

1. Take three pieces of mirrored perspex and tape them together to form a triangular shape. Make sure it is solid and the tape is on the outside of the triangle.
2. Trace around the small triangle at the end of the kaleidoscope onto the overhead transparency paper. (Add another 1cm all the way around the triangle to allow for folding).

3. Place the transparency paper onto the end of the kaleidoscope and cut slits at the corners so that the edges can be folded down.
4. Tape the transparency paper into place.
5. Draw another triangle, making this 2cm bigger than the last.
6. Decide what kind of coloured see-through plastic you would like to put inside your kaleidoscope. Cut out small pieces that will sit on top of the transparency paper.
7. Put the coloured plastic over the end of the kaleidoscope that has the transparency paper, and on top of that add the other (slightly bigger) triangle transparency paper. Tape the second triangle down on top so that there is still just enough room for the plastic to move between the two transparencies.
8. When your kaleidoscope is done, feel free to design and decorate a cover using cardboard, felt pens, glitter, tubing or anything attractive you want to use.

What's happening

The patterns inside your kaleidoscope are made by the light bouncing between the inside mirrors. While you look through one end, light enters through the other and reflects off the mirrors. Thus, varying colours and patterns are formed. Thanks to the symmetric pattern created by the well placed mirrors.

Make your Own Microscope

Make a simple microscope using water and take a closer look at the world around you.

The lens you create with water works like a microscope or magnifying glass, allowing you to see objects in much greater detail than if you were just looking with the naked eye.

What you need

A piece of fuse wire, Some water, Objects to look at (newspaper or a magazine with fine print works well.

What to do

1. Make a loop at the end of the fuse wire about 2mm wide.
2. Dip it into some water to get a drop formed in the loop.
3. Hold it close to your eyes and look closely at an object, such as a magazine.

4. You may have to experiment to get the right distance but you should see a magnified image, especially if you have the drop as close to your eyes as possible.

What's happening

Pioneers of early microscopes originally used tiny glass globes filled with water to magnify objects. This is similar to what you are doing in this experiment. The water droplet forms the shape of a convex lens, which refracts the light and converges it at the point where you see the image clearly. It was later that the method of grinding glass to make lenses was perfected. Modern microscopes have many lenses in them that allow us to see extremely small objects.

Check your Heart Rate

Make your own stethoscope and check your heart rate before comparing it to others. You might not be a real doctor but you can still use some of their medical equipment as part of this fun science project.

Check your heartbeat when resting and after physical activity, how fit are you?

What you need

A balloon, A piece of tubing, 2 small funnels, Scissors, A timer, Rubber band (optional) and A calculator (optional).

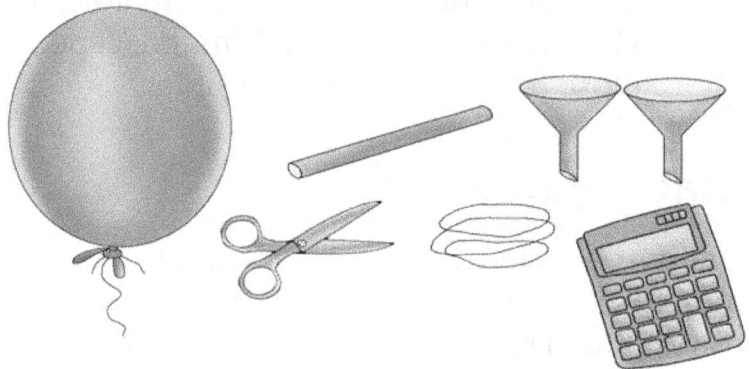

What to do

1. Take the piece of tubing and fit a funnel to each end.
2. Stretch the balloon by blowing it up and then letting the air out.
3. Cut off the top thread of the balloon with scissors.
4. Stretch the top thread of the balloon tightly over the open end of one the funnels. If necessary, use a rubber band to hold it in place.
5. Stir the mixture again with your fork and look at the long strands of gunk that have formed.
6. As the mixture cools slowly, add more water, small amounts at a time.

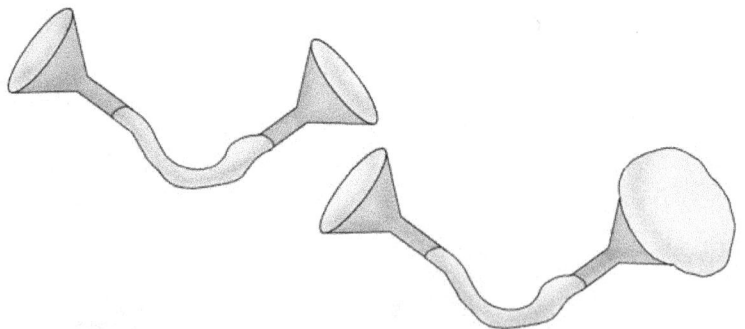

What makes your stethoscope work

1. Find your heart with your hand by feeling where it beats in your chest.
2. Sit down somewhere quiet and place the end of the funnel with the balloon over it against your chest, directly onto your skin, just to the left centre.
3. Hold the other funnel to your ear. You should hear a low beating sound.
4. Use the timer to count how many beats you hear in 20 seconds. Multiply this number by three (use a calculator if you're not confident) to find out how fast your heart beats in one minute.
5. Try doing some more tests, such as running around for 5 minutes and then checking how fast your heart is beating. Compare your results with your brothers, sisters, parents and even pets, heart rates. Are there any differences?

What's happening

Did you know that when a doctor listens to your heartbeat with a stethoscope, they are actually listening for two sounds? The first sound is a longer, lower pitched sound. The second is a shorter, higher pitched sound.

The lower pitched sound is made by the closing of two heart valves when blood is flowing out of the heart. The higher pitched sound is made by two other valves when blood is flowing into the heart. When a person exercises or participates in any kind of physical activity, the heart beats faster in order to pump more blood and oxygen to the muscles that are being used. The closing of the heart valves makes a sound which causes the stretched balloon to vibrate. The vibrating balloon makes the air in the tube vibrate and the tube then carries these sound vibrations to your ear.

Make a Model Hand

Working with other important parts of your body, muscles perform an important task.

Learn about how they work by making a model hand with working muscles and tendons.

What you need

A piece of card, the size of your hand, A pen or pencil, Scissors and A string.

What to do

1. Take the piece of card and trace the outline of your hand with a pen or pencil.
2. Cut out the shape of your hand with scissors.

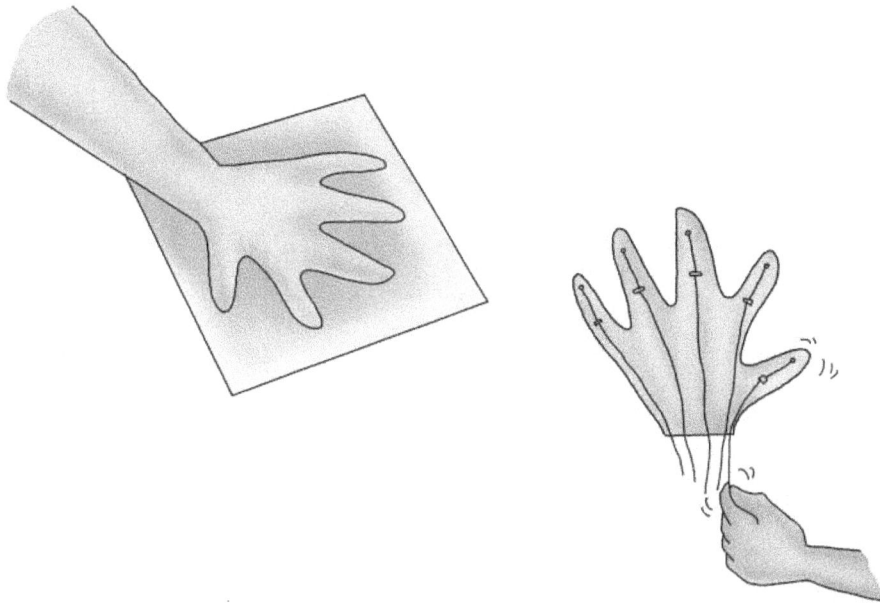

3. Cut the string into five pieces that are about the length of your hand.
4. Tie a piece of string to the tip of each finger and thumb, and stretch it to the base of the palm. Staple the string to the card at the same points, where you have joints in your fingers and thumb.
5. Try pulling the strings from the base of the palm, what happens?

What's happening

The muscles in your body are there to shorten or contract, a simple but very important task. Every movement you make is driven by the muscular system, from a simple smile to lifting a heavy box.

The muscles inside your forearm have long tendons running through ligament fibers, known as the *carpal tunnel*, in the wrist. These muscles allow you to flex your fingers, bending the tips towards your palm, as your fingers do when giving the thumbs up. This is what happens when you pull on the strings of your model hand.

Keeping Drinks Hot & More

Imagine you are making a hot drink. You are just about to add some milk when you realise that your neighbour is knocking at your door.

Should you add the milk now or wait until you have finished talking to your neighbour? Good question, time to do some research!

What you need

4 cups that are exactly of same size and same shape, Hot water, Cold water, Cold milk, A thermometer and Spoons.

What to do

1. Half fill each cup with hot water.
2. Check that all of the cups are at the same temperature. Leave the thermometer in one of the cups for now.

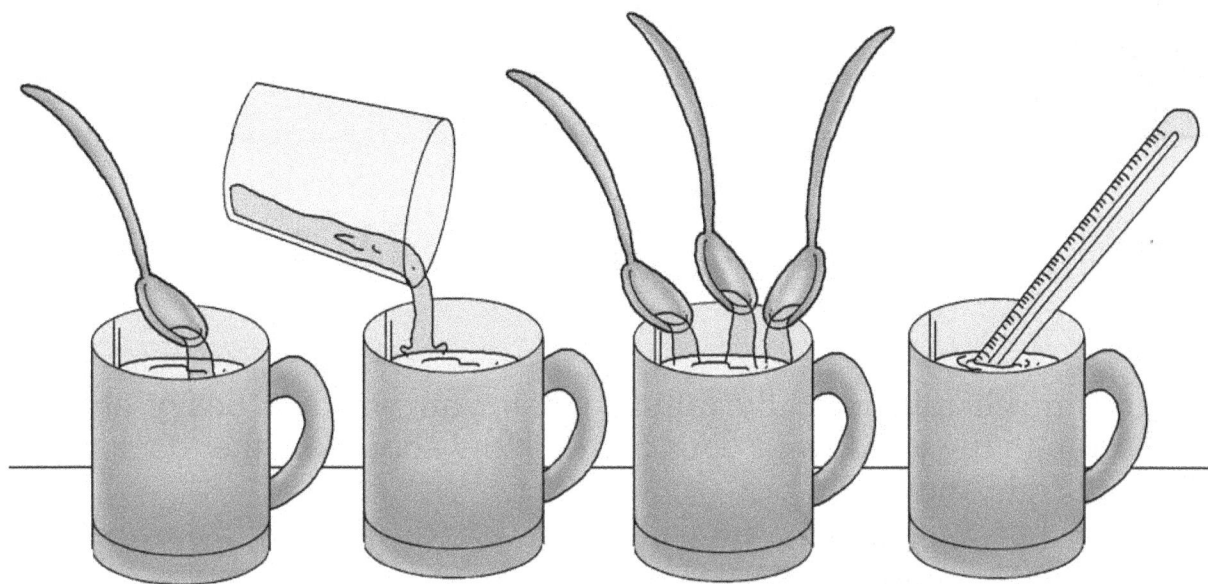

3. Add 1 spoon of milk to the first cup. Add 1 spoon of cold water to the second cup. Add 3 spoons of milk to the third cup. Don't add anything to the last cup.
4. Check the temperature of each cup every minute with the thermometer. Which cup of water stays hot for the longest?

What's more

This one is up to you, do you think you can explain it?

Some other interesting questions related to this topic include:

What are some good ways of keeping drinks hot?

Have you heard of the word, 'insulation'?

What happens if you want to keep a drink cool rather than hot?

Which is more likely to keep a drink hot for longer: a tall thin cup or a wide shallow cup?

Do some liquids cool faster than others?

What type of cup is better for keeping drinks hot: paper, plastic, clay or glass?

Make a Rain Gauge

How much rain is really falling when you watch a heavy shower through the window of your home? How about on other days when it's just a light shower?

Find out by making your own rain gauge, recording the results and studying your findings.

What you need

A plastic (soft drink) bottle, Some stones or pebbles, A tape, A marker (felt pen) and A ruler.

What to do

1. Cut the top off the bottle.
2. Place some stones at the bottom of the bottle. Turn the top upside down and tape it to the bottle.
3. Use a ruler and marker pen to make a scale on the bottle.

4. Pour water into the bottle until it reaches the bottom strip on the scale. Congratulations, you have finished your rain gauge.

5. Put your rain gauge outside when it starts raining. After a rain shower is through, check to see how far up the scale the water has risen.

What's happening

Rain falls into the top of the gauge and collects at the bottom, where it can be easily measured. Try comparing the amount of rain to the length of time the shower lasted. Was it a short and heavy rain shower or a long and light one?

If you want to get the exact figures, you can graph the rainfall over weeks or even months. This is especially interesting if the place you live experiences varying seasons, where sometimes it is very dry and other times, it is very wet.

Combine your results with the wind speed, wind direction and air pressure for a full weather report.

Make your Own Weather Vane

Knowing which direction the wind is blowing is an important yet often overlooked piece of information. Wind plays a role in many things we do and you'll know that is true if you've ever biked into a strong head wind, seen a forest fire, visited a wind farm or tried to predict weather changes.

Be aware of the wind direction by making your own wind vane and add it to your set of weather monitoring and predicting tools.

What you need

An ice cream container lid (or an old food container lid), Scissors or a craft knife (be careful and take help from parents and teachers when necessary), A marker (felt pen), A skewer, A straw and A pin.

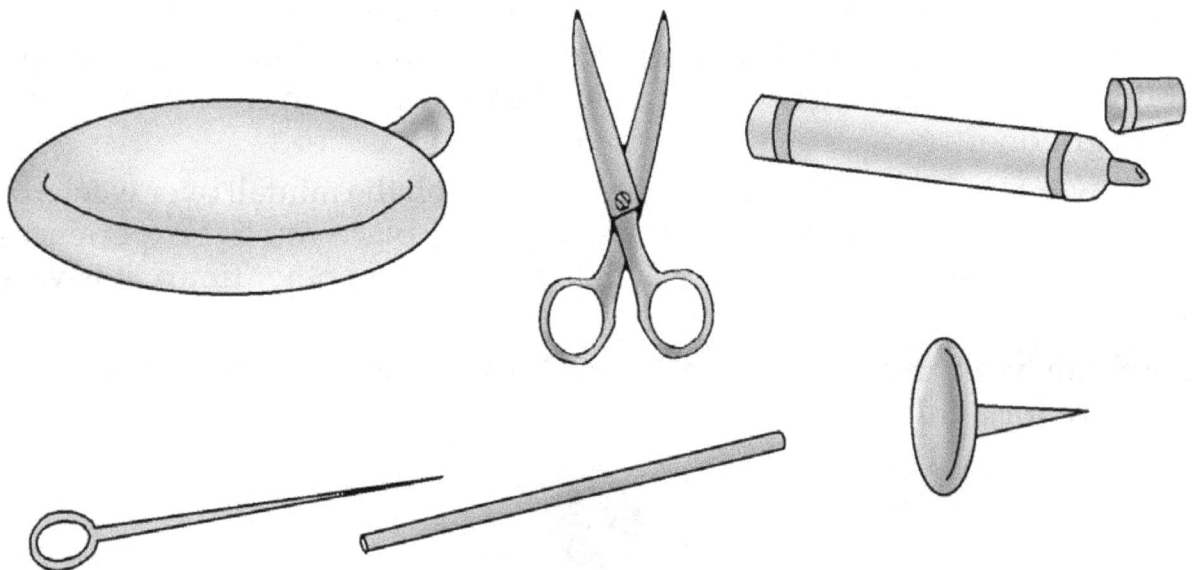

What to do

1. Trace a triangle onto the ice cream container lid with the marker and cut it out. Repeat the process, but this time trace and cut out a rectangle.
2. Cut a slit on both ends of the straw and slide the triangle in one end and the rectangle in the other end with a glue.
3. Push a pin through the exact middle of the straw and then into the flat end of the skewer.
4. Place it outside where you can easily see it from the inside and you'll be able to tell which direction the wind is blowing without going outside.

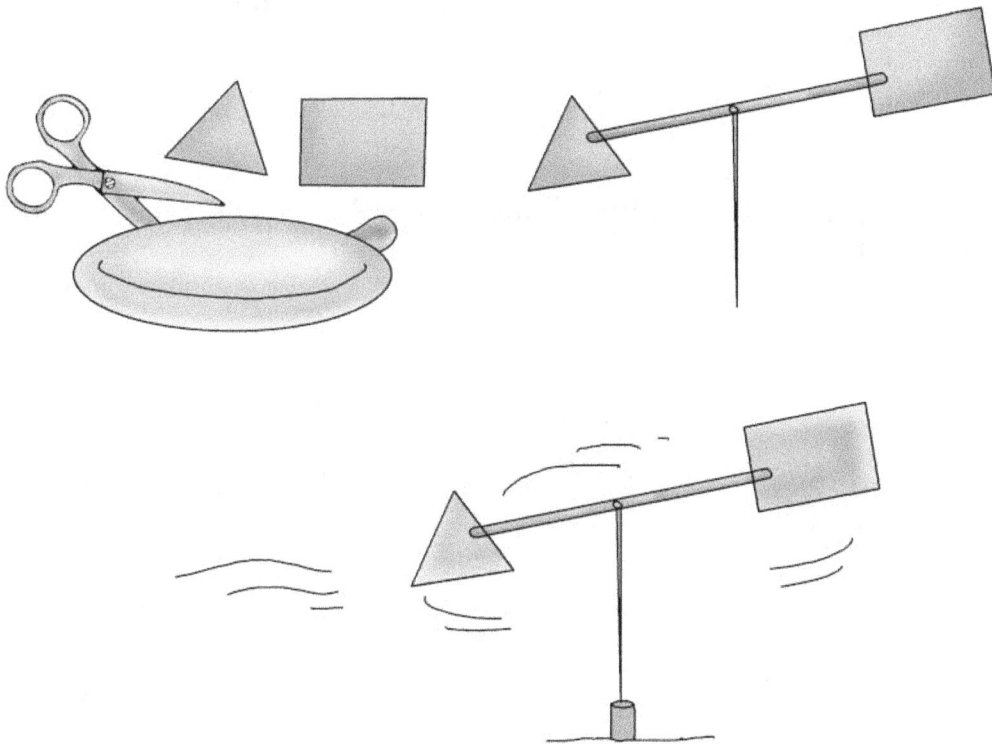

What's next?

Combine your wind direction results with the wind speed, air pressure and rainfall for a full weather report.

Make your Own Barometer

A barometer is used for measuring air pressure. It is a useful tool for predicting weather changes.

Make your own barometer and start making your own weather forecasts. Compare your results to the weather forecast on the news and see who does best!

What you need

A balloon, Scissors, A jar, A rubber band, Tape, A straw, A piece of card and A marker (felt pen).

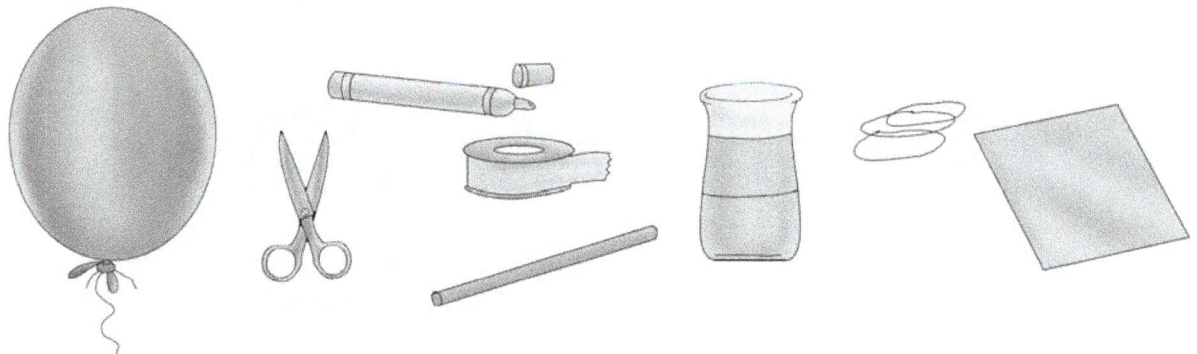

What to do

1. Cut the top of the balloon (the part which you blow into).
2. Stretch the balloon over the top of the jar and hold it in place with a rubber band.
3. Keep the straw across the top of the jar so that one-third of the straw is hanging over the edge. Stick the straw to the balloon with the tape.
4. Draw three lines on the piece of card that are about half a centimetre apart from each other. Label these lines as *high*, *moderate* and *low*.

5. Tape the card against the back of the jar so that the straw points to the moderate line.
6. Put your barometer on a flat surface somewhere inside.

What's happening

When there is low air pressure, the balloon should expand out and the straw will point down. This is because the air inside the balloon now has relatively more air pressure compared to the air outside. It pushes the balloon out as a result.

When there is high air pressure, the air on the outside will push the balloon into the jar and the straw will point upwards. The air inside the balloon now has relatively less pressure. This pushes the balloon inwards as a result.

In general, high air pressure indicates fair weather, while low air pressure indicates that bad weather is more likely. Although forecasting the weather isn't an exact science and can be very difficult at times, give it a try and see how accurate you are.

Combine your results with wind speed, wind direction and rainfall for a full weather report.

Take the Wind Speed Challenge — Anemometer

Harness the power of wind with some weather based projects that will help you understand wind speed and how it changes from place to place and day to day.

Put your problem solving skills to the test with these fun challenges.

An anemometer will help show you how fast the wind is blaming spinning the cups around. The faster the wind is blowing, the faster the cups will spin.

What you need

Paper cups, A skewer (or something similar to poke holes), Straws, Scissors, A marker (felt pen) and A tape or glue.

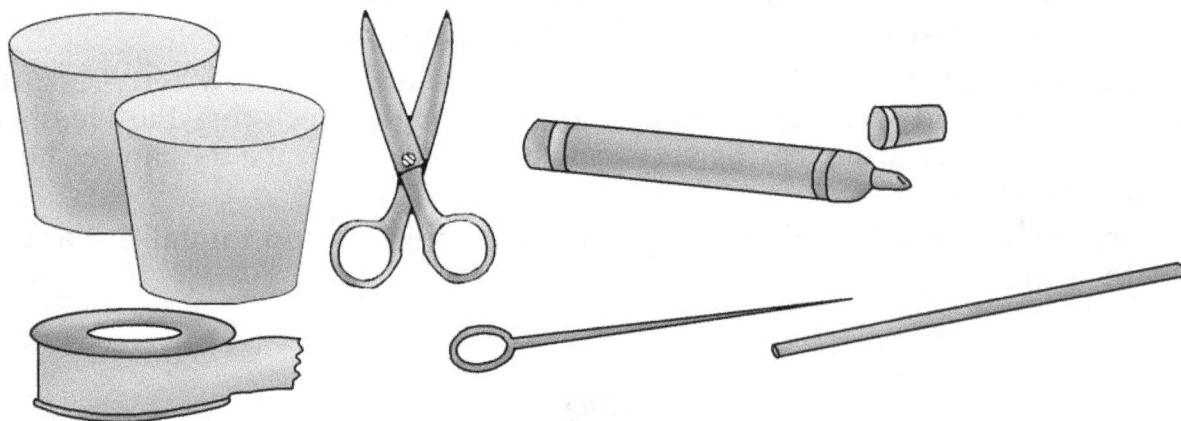

What's the challenge

Your challenge is to design something that can measure the wind speed. Create an anemometer that features free spinning cups that spin faster as the wind increases. The wind should blow into the cups pushing them away. The faster the wind, the more force it has to push the cups and the faster they spin. You can measure the wind strength by comparing how many times the anemometer spins around every 10 seconds. Does it vary from place to place and day to day?

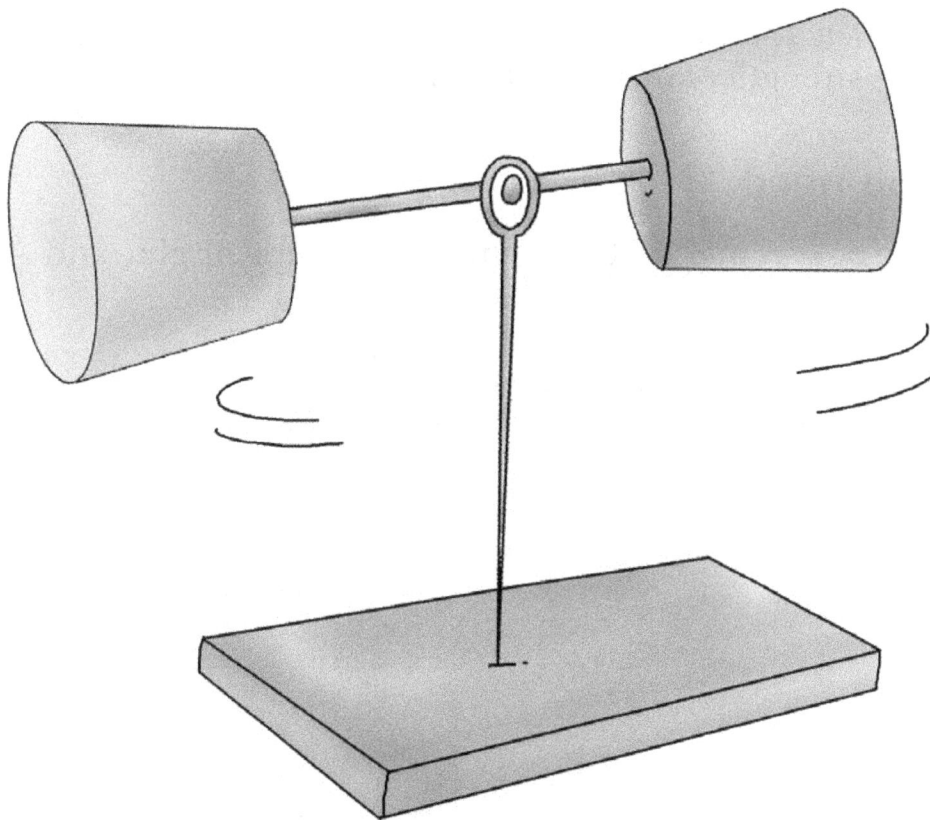

Wind Speed Box

Make a wind speed box to measure how fast the wind is blowing. Similar in use to an anemometer, your wind box will be able to measure the strength of the wind in different places.

What you need

An old shoe box, A marker (felt pen), String, Tape or glue and A piece of card.

What's the challenge

The challenge is to make a wind box that can compare different wind speeds. Marking a scale inside the box is a good place to start with, and you can use the piece of card as a guide, with it swinging further along the scale as the wind increases. The rest is up to you and your problem solving skills. A stronger wind has more force to push the piece of card along the scale, while it might struggle to move it at all on a very calm day.

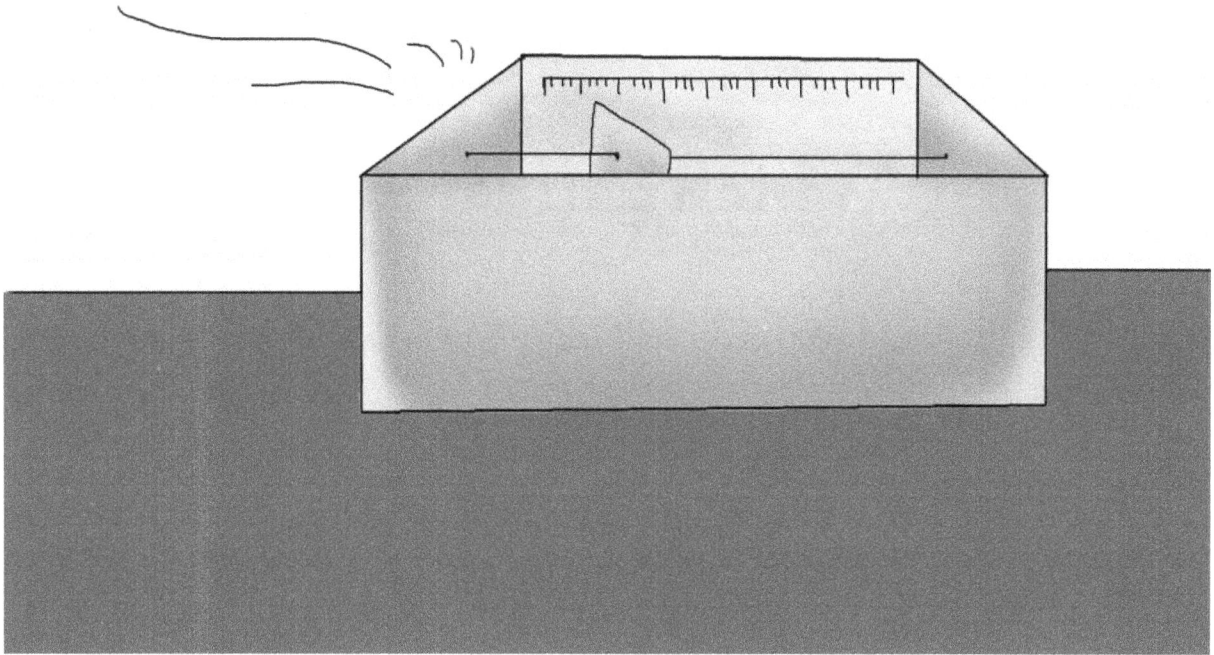

What's next

Combine your wind speed results with wind direction, air pressure and rainfall for a full weather report.

Which will Burn out First?

In this simple experiment, you will learn how a candle needs oxygen to burn.

What you need

3 candles, 3 saucers, 1 small jar and 1 large jar.

Note:
This experiment requires adult supervision.

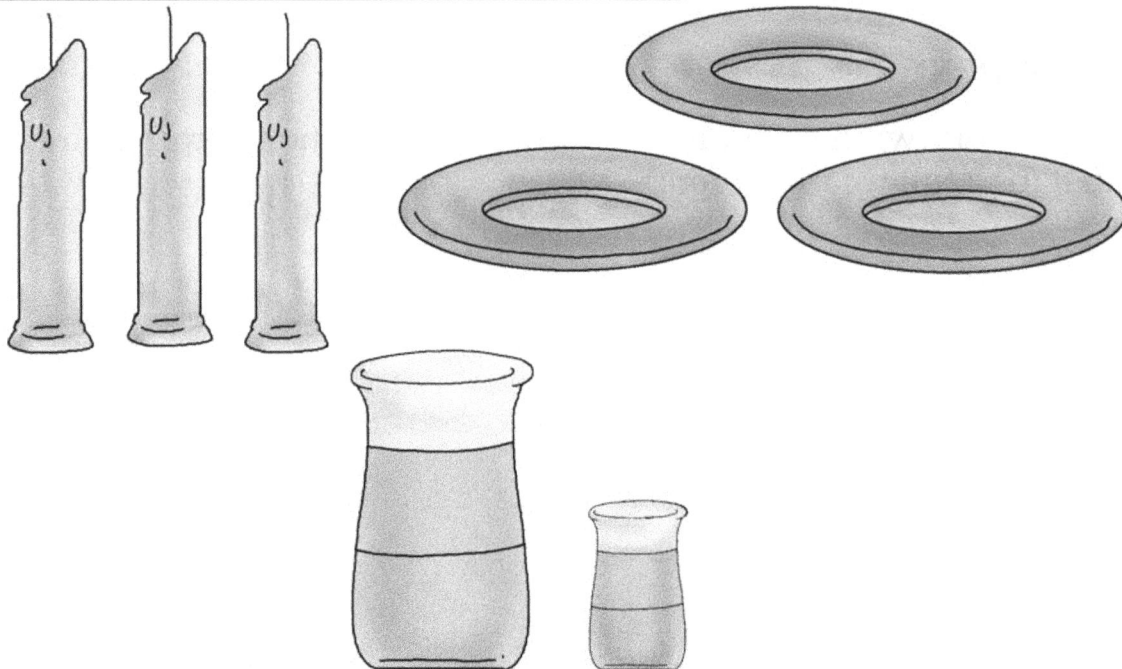

What to do

1. Place each of the three candles firmly on a saucer.

2. Light all the three candles.
3. Leave the first candle out in open air.
4. Place the small jar over the second candle.
5. Place the large jar over the third candle.
6. Which burns out first? Which burns out second? Which lasts the longest?

What's happening

The candle in the small jar burns out first because it has the least oxygen to burn. The candle in the large jar burns out next, because although it had more oxygen to burn than the candle in the small jar, it still runs out of oxygen. The candle in the open air continues to burn the longest, because it does not have a limited the amount of oxygen to burn.

Note:

Be careful when removing the jars, they could be hot.

Surface Tension

Did you know that water molecules are attracted to each other?

What you need

A Glass of water, Dry paintbrush.

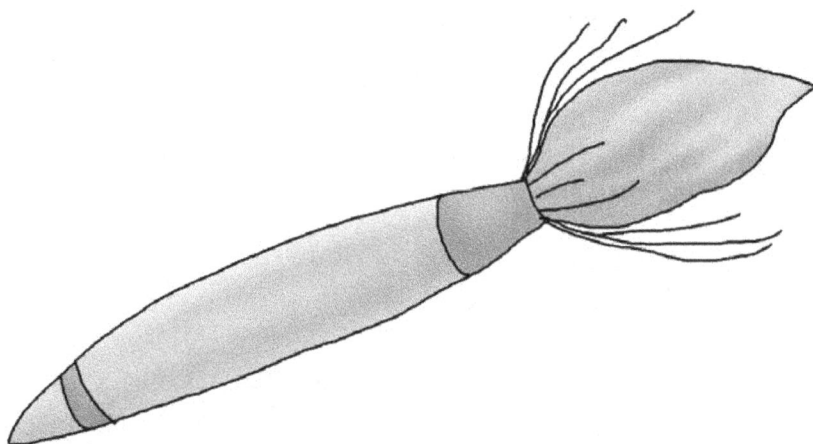

What to do

1. Make sure that the paintbrush is dry.
2. Notice how the bristles of the paintbrush look while they are dry.
3. Now dip the tip of the brush into the water, and notice how the bristles are all pulled together at the tip as they are wet now.
4. The *Surface Tension* of water is strong, and pulls together the bristles and the water.

What's happening

The property of a liquid surface causes it to act like a stretched elastic membrane. Its strength depends on the forces of attraction among the particles of the liquid itself and with the particles of the gas, solid, or liquid with which it comes in contact. The surface tension allows certain bristles to stand on the surface of water and can support a razor blade placed horizontally on the liquid's surface, even though the blade may be denser than the liquid and unable to float. Surface tension results in spherical drops of liquid, as the liquid tends to minimise its surface area.

Gumdrop Architecture

Not only eating gumdrops is fun, but they can also be used in making great things. With them you can build a bridge or a tall tower!

What you need

A bag of gumdrops and A box of round toothpicks.

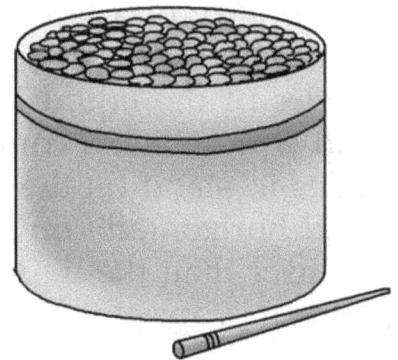

What to do

1. Start out with a foundation of gumdrops, and build up from there!
2. Use the toothpicks to poke into the gumdrops at each intersection of toothpicks, to hold the structure together.
3. Make squares, cubes, pyramids, triangles, or anything you like!

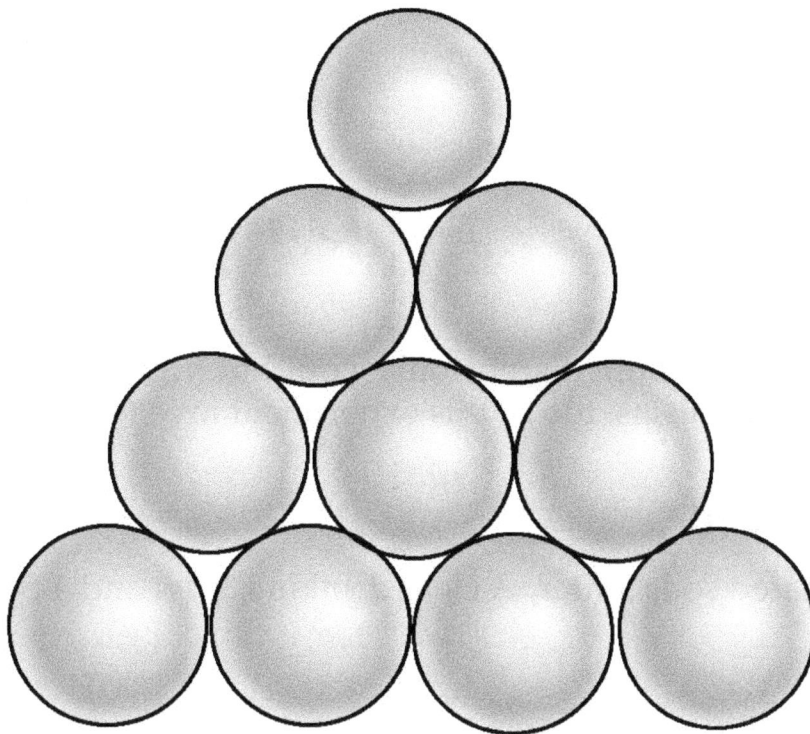

What's more

Instead of gumdrops, you can use balls of clay or mini marshmallows. Each of these will dry to make a solid structure.

Note:

Which shapes make stronger structures? Make a bridge out of different shapes and see which one can hold heavier objects. This could make a good and interesting science project!

Dancing Raisins

Do you love dancing? It's not only you, but you will be amazed to discover that raisins, too in this experiment love to dance!

What you need

Clear, carbonated soda water, A clear drinking glass and 4-5 raisins.

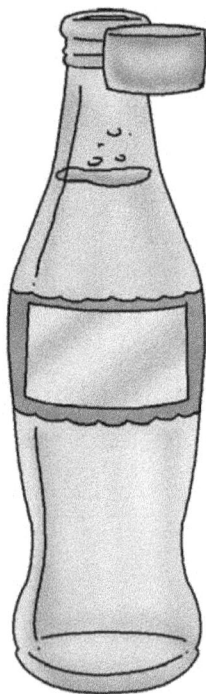

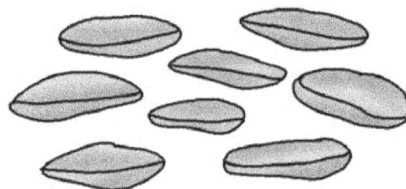

What to do

1. Pour clear carbonated soda water into a clear glass.

2. Drop four or five raisins into the glass.
3. After about a minute, you will observe the raisins moving up and down in the glass. Watch how the bubbles control the movements of the raisins! Wonderful, isn't it?

An Egg in a Bottle

How can you put an egg inside a bottle or a narrow-mouthed glass? Try doing the experiment and find it out all by yourself!

What you need

An empty glass or an apple cider container or an apple juice jug, Newspapers, Hard boiled eggs and Matches.

Note:

This experiment requires adult supervision!

What to do

1. You need an empty glass or an apple cider container or an apple juice jug.

2. Put torn newspaper pieces in the jug and light it up. Please use proper safety precautions.
3. As soon as the fire gets going, place a hard boiled egg on the top of the jug.
4. Soon the egg will get sucked inside into the jar with a pop.

What's happening

Actually, the burning newspaper pieces heats the air trapped inside the bottle causing it to expand. After a short while, the fire inside the bottle dies, thus causing the air inside to cool down resulting in a low pressure inside the bottle. The egg is forced/sucked into the bottle because there is a lower pressure inside and a greater pressure outside the bottle.

Note:

The container you use should have an opening only slightly smaller than the egg.

Balloon in a Bottle

Watch a bottle inhale a balloon!

What you need

A empty soda or water bottle, A balloon, A large bowl, Hot water and Ice water.

What to do

1. Fill the bottle with hot water, swirl the water around to make the bottle hot, and pour it out.
2. Refill the bottle 1/4th full with hot water and place the balloon on the top.
3. Now, fill the bowl with ice water, and place the bottle in the bowl.
4. Watch as all of the air is taken out from the balloon. It might even get pulled into the bottle!

What's happening

This works because hot air expands and cool air contracts. So when you first put the balloon on the top of the bottle, the air inside the bottle is hot. As the air cools from the ice water, it contracts and tries to pull more air in from the outside.

Clickety-Clackety Coin

Learn how to make a coin move without touching it.

What you need

2-litre plastic soda bottle, A quarter glass of Water and a Coin.

What to do

1. Place the empty uncapped bottle in the freezer for 10 minutes.
2. Dip the coin in water.
3. Take out the bottle from the freezer and immediately place the wet coin on the top of the open bottle.
4. Wait and watch the coin move! As it moves, it will make a clickety-clackety noise.

What's happening

When the bottle was taken out from the freezer, the cold air inside the bottle expanded and tried to rush out of the bottle. This air flow caused the coin to move!

Alka-Seltzer Rockets

Take a trip to the moon with these gas blasters!

What you need

1/4th of an Alka-Seltzer tablet, 1 Fuji film canister – the kind where the lid fits inside the canister and Warm water.

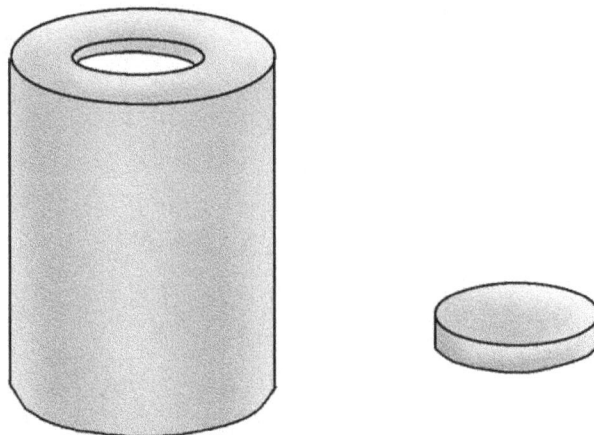

Note:

Be sure to do this experiment outside and with adult supervision.

What to do

1. Fill the film canister 1/2 full with warm water.
2. Drop in 1/4th of the tablet of Alka-Seltzer. Do not use more than the amount mentioned.
3. Snap the lid tightly into the canister, turn it over on a hard surface, and stand back! The sodium bicarbonate ($NaHCO_3$) will make the canister launch into the air. If you vary the temperature of the

water, your rocket will shoot to different heights. The warmer the water, the higher it will go.

4. If you want to make a rocket launch pad, cut three slits about 1 inch high in the bottom of a toilet paper tube. Bend the slits so you can tape them to a paper plate. You might want to decorate your plate and tube first.

5. When you are ready to launch your rocket, drop the prepared film canister into the toilet paper tube and stand back! It goes up with a boom as shown in the above figure.

What's happening

The hot gases provided by combustion are ejected in a jet through a nozzle at the rear of the rocket. The term is also commonly applied to any of the various vehicles, including fireworks, skyrockets, guided missiles, and launch vehicles for spacecraft, that are driven by such a propulsive device. Typically, thrust (force causing forward motion) is produced by reaction to a backward (rearward) expulsion of hot gases at extremely high speed.

Speedboat Matchsticks

Watch a matchstick zoom across the water's surface.

What you need

A bowl of water, Wooden matchsticks, Liquid dish soap and some Sugar cubes.

SOAP

What to do

1. Put a few matchsticks in a bowl of water.
2. Drop a small amount of dish soap into the centre of the bowl and watch the matchstick shoot across the surface of the water like power boats!
3. Clean out the bowl and refill it with water and the matchsticks.
4. Lower sugar cube into the bowl and watch the matchsticks float towards it!

What's happening

The soap gives off an oily film that rushes outwards and pushes the matchsticks away, while the sugar cube is porous and sucks water inwards, pulling the matchsticks towards it.

Making A Speedboat

Turn an empty soda bottle into a speedboat.

What you need

2 litre soda bottle, Straw, Clay, Paper napkins, 1/2 tablespoon of baking soda, 1/4th cup of vinegar and a Straw.

What to do

1. Cut a small hole in the bottom of your soda bottle, and poke the straw half way through.
2. Plug the gap around the straw with clay so that no air can escape out of the bottom of the bottle except through the straw.

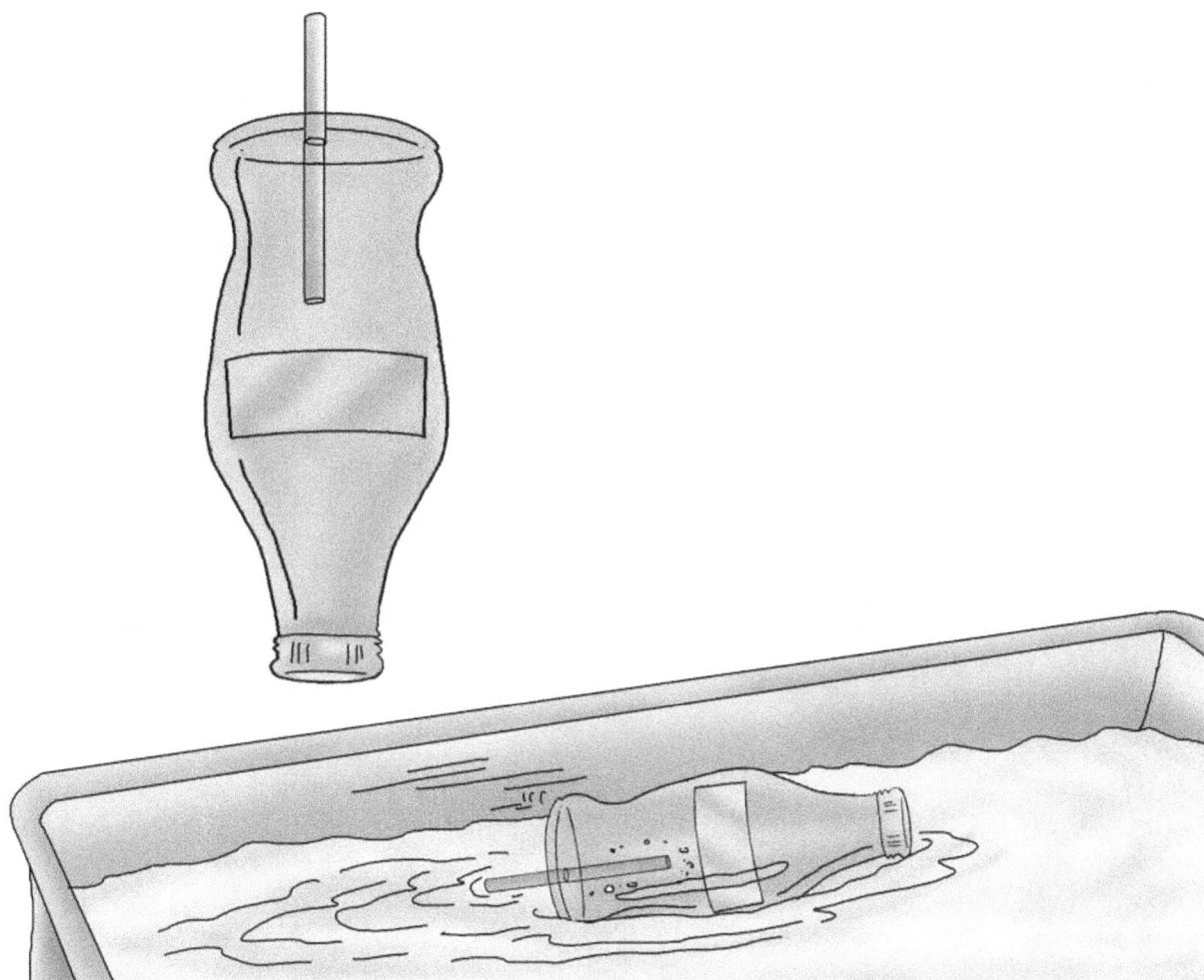

3. Put the baking soda in the centre of your paper napkins, fold it lengthwise and twist the ends closed. This is to protect the baking soda from the vinegar for a few seconds.

4. Pour the vinegar into the bottle, add the paper napkin with the baking soda and put the cap on quickly.

5. Put the bottle into the water tub (pool or bathtub) with the straw submerged under the water surface like a motor.

6. As the *baking soda* and *vinegar* begin to react, the "boat" will be powered forward! Thus your speedboat is ready!

Blow it Up

Blow up a balloon without having it touch your lips.

What you need

A empty soda or water bottle, A Balloon, Baking Soda, Vinegar, Paper towel or Paper napkin.

What to do

1. Tear the paper towel in half. Take one half and tear it in half again. You will use one of these smaller squares as your wrapper.
2. Place about a tablespoon of baking soda on your wrapper. Fold it up and twist the ends closed so that the baking soda is neatly inside.

3. Pour about 1/4 cup of vinegar into your bottle, and add the wrapper of baking soda.
4. Quickly put the balloon securely over the top of the bottle and watch the balloon blow up by itself!

What's happening

You can swirl the liquid in the bottle to help the two chemicals react, once the balloon is fastened to the top.

This experiment works because the vinegar and baking soda get mixed together to produce a gas which fills the bottle and the balloon. The paper towel is used to protect the baking soda for a short period of time while the balloon gets placed on the bottle.

Oil Blob Dance

Watch an oil blob dance between two different liquids!

What you need

A clear drinking glass or plastic cup, 1/2 cup of water, 1/2 cup of rubbing alcohol, Cooking oil and an Eyedropper.

What to do

Watch an oil blob dance between two different liquids! Here's what to do:

1. Pour the water into the glass.
2. Tilt the glass slightly and slowly pour in the rubbing alcohol. Don't shake the glass, or the two liquids will mix.
3. Fill the eyedropper with the cooking oil and lower the tip into the layer of rubbing alcohol, but not into the water.

4. Squeeze out a couple of drops of the oil, and watch the oil blobs dance!

What's happening

The oil is lighter than the water but heavier than the rubbing alcohol. So the *oil blobs float* between the two liquids.

Note:
Use adult supervision with the rubbing alcohol.

Build your Own Electromagnet!

Try making this special kind of magnet at home from things you may already have around your house.

What you need

A large iron nail (about 3 inches), About 3 feet of thin coated copper wire, A fresh D size battery and Some paper clips.

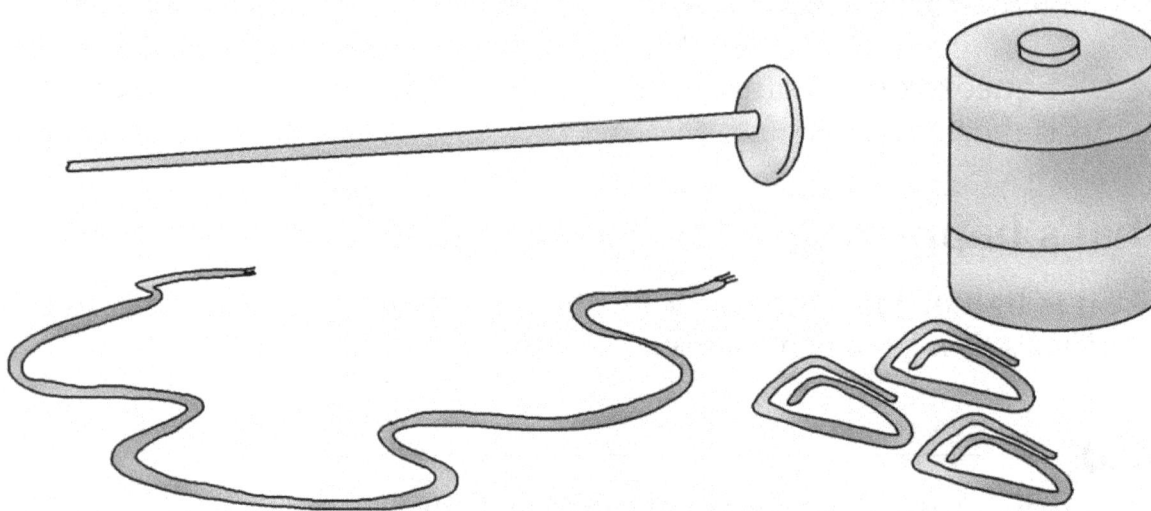

What to do

1. Leave about 8 inches of wire loose at one end and wrap most of the rest of the wire around the nail. Try not to overlap the wires.
2. Cut the wire (if needed) so that there is about another 8 inches loose at the other end too.

3. Now remove about an inch of the plastic coating from both ends of the wire and attach the one wire to one end of a battery and the other wire to the other end of the battery. (It is best to tape the wires to the battery - be careful though, the wire could get very hot!)

What's happening

Now you have an *Electromagnet*! Put the point of the nail near a few paper clips and it should pick them up! Most magnets, like the ones on many refrigerators, cannot be turned off, and they are called *permanent magnets*. Magnets like the one you made that can be turned on and off, are called Electromagnets. They run on *electricity* and are only magnetic when the electricity is flowing. The electricity flowing through the wire arranges the molecules in the nail so that they are attracted to certain metals. Never get the wires of the electromagnet near or at household outlet! Be safe and have fun!

A Duck Call

Use a plastic straw to make a duck call!

What you need

One plastic straw from your kitchen or local fast food restaurant, Scissors, Lungs (don't worry you already have them to blow air into the straw.)

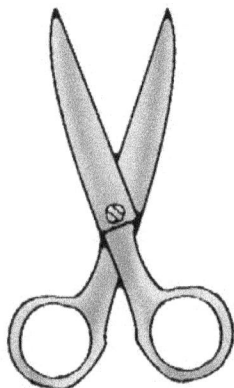

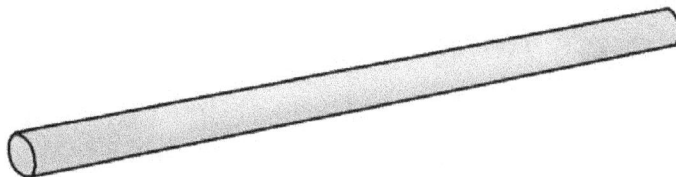

What to do

1. Use your fingers to press on one end of the straw to flatten it – the flatter the better.
2. Cut the flattened end of the straw into a point as shown in the figure.
3. Flatten it out again real good.
4. Now take a deep breath, put the pointed end of the straw in your mouth and blow hard into the straw. If all goes well, you should

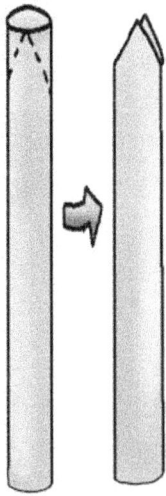

hear a somewhat silly sound coming from the straw. The smaller you are, the harder it may be to get a good sound. Sometimes adults can get more sound because of their bigger lungs. If you still have trouble, try flattening it out some more.

5. Don't stop there; try cutting the straw in half to see how the sound changes, or make another identical straw and add the pointed end of the new straw to the uncut end of the first straw to make the first straw longer. The sound will be very different, (more like a moose call!) and you will have to blow even harder, but give it a try.

What's happening

This is science? It sure is. You see all *sounds* come from *vibrations*. That little triangle that you cut in the straw forced the two pieces of the point to vibrate very fast against each other when you blew through the straw. Those vibrations from your breath going through the straw created that strange duck-like sound that you heard. Now you will never be bored again when you go to a fast food restaurant! Have fun but watch out not to disturb anyone.

71+10 New Science Projects Self-learning Kit

— Dr. C.L. Garg/Amit Garg

(Also available in Bangla, Tamil)

Science projects and models play a pivotal role in inculcating scientific temper in young minds and in harnessing their skills. Students of classes 10 th, 11th & 12 th have to work on such projects and these carry much weight in the overall performance.

All these aspects have been considered during the compilation of the projects and models. This book will also be an ideal choice for parents interested in enhancing scientific temper of their children and for hobbyists.

81 Classroom projects on: Physics, Chemistry, Biology& Electronics for Sec. & Sr. Sec. Students

Big Size • Page: 144
₹ 140/-(with CD) • Postage: ₹ 20/-

Drawing & Painting Course

–A.H. Hashmi

Children have always been attracted towards bright colours, various shapes and diverse objects that they see around them. Nature fascinates them. The beautiful birds, animals, flowers and trees fire their imagination and they want to capture it on paper. But how, for all are not artists.

Well, this book has been especially developed for those who want to learn and master the art in a fun way. The step-by-step instructions, along with the audio-visual CD, will show you how to create beautiful pictures. See how a circle or an oval transforms into a flower or a peacock; a few lines here, and a few there become a human figure.

This book starts with the basics Ã,Â— lines, shades, texture, balance, harmony, rhythm, tone, colours, etc. and goes on to teach the different techniques of drawing and painting.

So pick up a pencil and paper and let your imagination fly. Gain confidence with each passing day and master the art of drawing and painting.

Big Size • Page: 124
₹ 150/- • Postage: ₹ 20/-

101
Science Games

– Ivar Utial

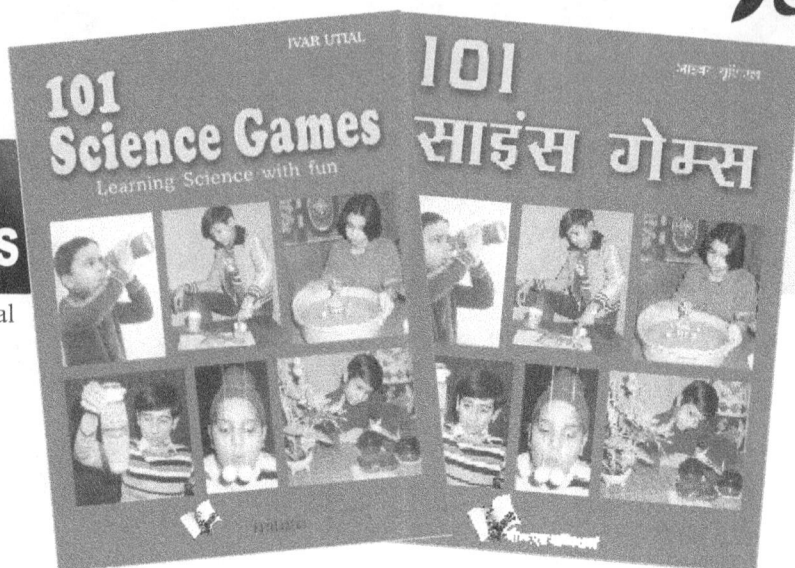

Supplementary science books not only interest and excite young students, but also stimulate their interest in the subject.

This exciting book shows you how to have fun with 101 Science Games. There is little doubt that science experiments can be quite interesting and useful in discovering mysteries of nature. In fact, it is the science that has led man from the lantern and the bullock-cart age to Electronics and Supersonic Jet age.

The book is fully illustrated with step-by-step instructions to give you first hand experience of making simple scientific equipments like :

- Telescope
- Barometer
- Hectometer
- Model Electric Motor
- Electroscope
- Periscope
- Steam Turbine

This project-work will acquaint you practically with the basic principles of specific applications. Now, you too can prepare for your next science fair.

Big Size • Page: 120
₹ 96/- • Postage: ₹ 20/-

Greatest
Crafts & Projects for Children

– Vikas Khatri

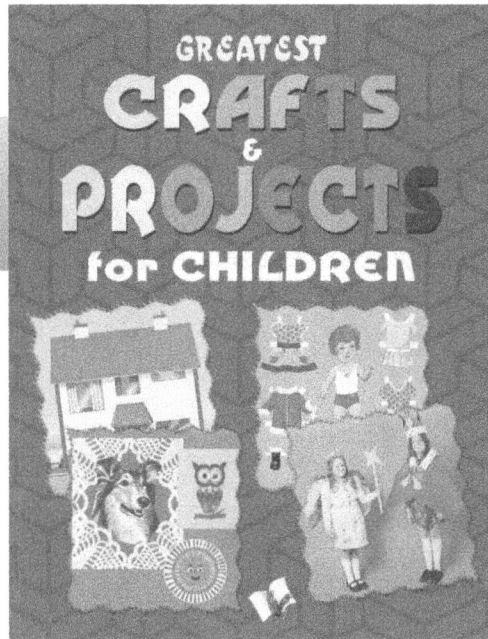

54 cool and awesome projects, crafts, experiments and more for kids!!!

Unplugging kids from their MP3 players and game systems for one-on-one family time is a great way to reconnect in today's hectic world. And what better way to spend time together than doing an activity that's not only fun but also promotes creativity and self-expression?

Greatest Crafts and Projects for Children is packed with 54 craft projects ranging from outdoor projects to gifts and party favors to holiday decor to projects that promote learning through play with step-by-step instructions to guide children to successful completion of each project.

Filled with easy-to-follow instructions and fun, Greatest Crafts and Projects for Children provides parents, caregivers, and teachers with the tools they need to make the children recognise and cultivate the creativity within themselves along with fun.

Big Size • Page: 112
₹ 100/- • Postage: ₹ 20/-

The Portrait of a Super Student

– Abhishek Thakore

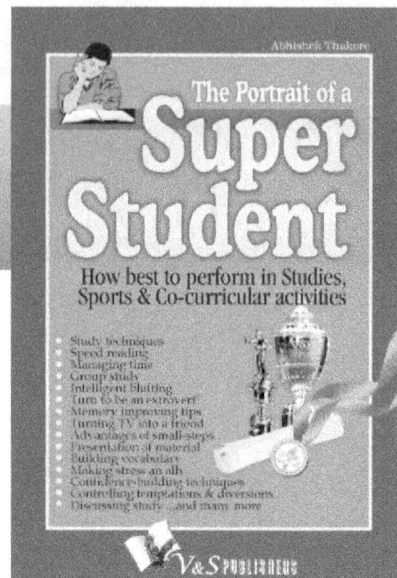

Success today depends a lot on one's academic achievements. And to excel in studies, you don't have to be just an intelligent or brilliant student — but also one who knows how to manage studies and time. In fact, even a mediocre or a below-average student can perform exceedingly well by following a scientific system.

The Portrait of a Super Student now brings you an innovative system, specifically designed for super achievement. From simple, practical and time-tested tips on how to manage time, controlling temptation, scheduling time and work, relaxing techniques to diet control, speed reading, building vocabulary, improving presentation, discussing studies —it goes on to guide you on how to make stress an ally, make a friend out of your TV. And above all, to make it reader-friendly the book is divided into easy-to-read small chapters —with a practice section after every chapter.

Demy Size • Page: 152
₹ 110/- • Postage: ₹ 20/-

7 Mantras
to Excel in Exams

– Prem P. Bhalla

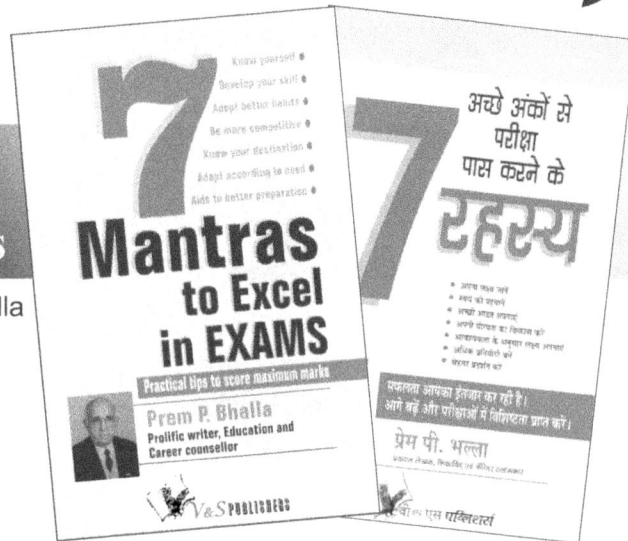

Exams play a major role not just during acadenic pursuits, but later in the career too. Although youngsters are taught a variety of subjects to equip them for life in general, no school teaches them how to excel in exams. Most learn only through trial and error. Others remain clueless about how to excel in exams throughout their lives. But this crucial information ensures that even those with average IQ can excel in exams. This book contains simple and practical tips and guidelines on how to tap your full potential and give off your best during exams. It is an invaluable guide for all students and adults due to appear in exams, as well as for parents who wish to ensure that their children do well and secure maximum marks.

The book offers simple guidelines on:
- Improving memory
- Maximising concentration
- Adopting effective study habits and techniques
- Developing proper reading, listening, language and communication skills
- Doing well in different kinds of exams
- Understanding what the examiner wants
- Overcoming exam anxiety and tension

Size • Page: 144
₹ 80/- • Postage: ₹ 20/-

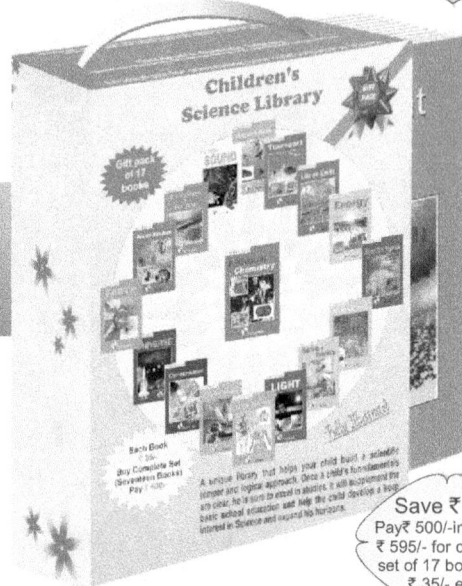

A Youngsters' Guide to
PERSONALITY DEVELOPMENT

– S.P.Sharma

A Youngsters' Guide to
PERSONALITY DEVELOPMENT

S.P. Sharma

A Book for Young Men & Women, especially
Students, with Indian Precepts & Culture

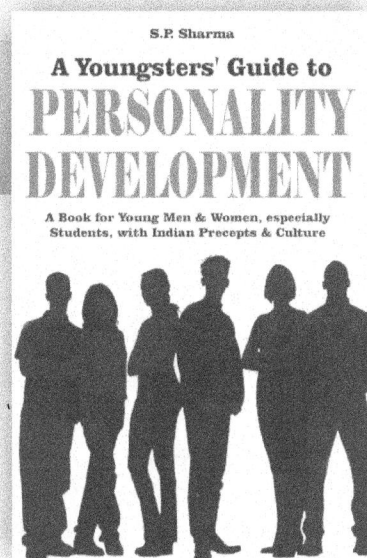

Acquiring a good personality is the most desirable attribute everyone aspires for. It acts as a catapult for a bright and successful life. The guidance of parents and teachers apart, an individual has to make the best use of his experience to develop and refine his own personality in a balanced way.

Highlights:

➢ Appreciate that knowledge is power. There is no age limit for learning, unlearning and relearning.

➢ Identify persons of high excellence as role models and follow their life sketch. Develop attitudes free, to the extent possible, from prejudices and biases.

➢ Inculcate several skills. For any profession or job, it pays to prepare well for the examinations, to be proficient in more than one skill-set, and acquire positive outlook.

This book is easy to understand, and has plenty of examples. It will be found very useful for effectively guiding the young students, who are on the threshold of embarking on a new life. Since the book derives substance from Indian ethos and culture, assimilating them into personality should be easy.

An indispensable book for the young students!

Size • Page: 120
₹ 110/- • Postage: ₹ 20/-

www.ingramcontent.com/pod-product-compliance
Lightning Source LLC
Chambersburg PA
CBHW080555220326
41599CB00032B/6483